HOW TO NOT F*CK UP A CRM SYSTEM

How To Not F*ck Up Series

MICHAEL M HUDLOW

Michael M. Hudlow

ISBN 978-1-7360213-6-1

Published by: Michael M. Hudlow

Email: Publisher@hownottofckup.com

Originally published in August 2023

Edition 1, Version 2

☮️,🩶,😁

Contents

Part I: Evaluating and Pre-Work

Part II: Implementing And Going Through The Setup

Part III: Optimizing The Operations

How Not To F*ck Up A CRM

How Not To
F*CK UP
A CRM
SYSTEM

Michael M Hudlow

Authors Forward

Based Upon Reality, Not Theory

In this book "How Not To F*ck Up A CRM System," I share real-life stories and offer invaluable insights to help you create a successful CRM implementation. One key focus is satisfying people and ensuring their contributions to the system are effortless while providing them significant value in return. The goal is to establish a process that minimizes the data entry burden, making it far outweigh the benefits they receive from the system. It is even better if their specific contributions can directly correlate with increased business value, such as generating more sales, accelerating business development, and boosting personal commission.

Throughout the book, I dispel the myth of finding a silver bullet solution that magically makes everyone happy. The truth is, there is no perfect company to install and operate the CRM system. The idea of an ideal installation partner is also just a myth. And most importantly, you will never have a perfect set of users. Instead, you will encounter a diverse mix of individuals, ranging from those eager to help, understanding, and committed individuals to those who are time-stretched, unenthusiastic, overwhelmed, or even obstructive.

To navigate these challenges, the book emphasizes the need to change your behavior and set realistic goals and expectations for system usage and the corresponding value you expect to achieve. It is crucial to adapt to and manage the different attitudes and behaviors of the users, ensuring they understand the benefits of the CRM system and fostering a collaborative environment.

Drawing on real-world case studies, "How Not To F*ck Up A CRM System" provides you with practical guidance and insights into scenarios that other organizations have faced. By learning from their experiences, you can make better choices for your own company, avoid common pitfalls, and ensure a successful CRM implementation that drives meaningful business outcomes.

In summary, this book serves as a mentor, offering practical advice and wisdom garnered from extensive hands-on experience. It highlights the importance of focusing on user satisfaction, dispelling myths, adapting behavior, setting realistic goals, and leveraging real-world case studies to guide you through the challenges of implementing a CRM system.

The Issues This New Tech Created

As a seasoned CRM professional, I understand many C-level executives' fear and skepticism when implementing a new CRM system. The revolutionary ideas and process changes involved can make even the most tech-savvy professionals feel overwhelmed and out of their depth. Many may assume that the changes only apply to younger, more technologically advanced companies and that their old-school methods of paper Rolodexes, Blackberries, fax machines, and typewriters will suffice. However, failing to adopt a modern CRM system can significantly hinder business success and the ability to remain competitive.

That's where How Not To F*ck Up A CRM comes in, a book designed to help you avoid the pitfalls and mistakes commonly made while adopting and implementing new CRM technologies. While purchasing a modern CRM system may be relatively easy, implementing it successfully is another issue. Mismanagement and misuse can result in a system that operates at a fraction of its potential capacity, causing frustration, unnecessary costs, and lost opportunities.

This book will guide you through real-life scenarios and stories that challenge traditional advice often given by project managers or CRM implementers. The anecdotes presented here come from countless conversations and examinations of serious situations that may seem unreal but have indeed happened to others. These stories are anonymized to protect the people involved, but they are accurate to the core, allowing you to gain insights from the experiences of others.

In How Not To F*ck Up A CRM, I present scenarios, warning signs, and lessons learned to help you avoid the same mistakes. The book provides practical solutions to common problems encountered before, during, and after implementation. Additionally, I advise spotting lurking issues, challenging decisions, and recommending alternatives.

This book is not a replacement for professional assistance and services but rather a tool to help prepare you personally for the challenging process of adopting a new CRM system. With the right mindset and approach, I assure you that you can successfully navigate the implementation process and reap the benefits of a modern CRM system.

Throughout the book, I recount my personal experiences, the catastrophic adventures I've had, and the valuable lessons I've learned as I embarked on a learning journey through case studies. With How Not To F*ck Up A CRM, I hope to be your mentor and guide, helping you avoid the mistakes others have made in pursuit of CRM success. You can create a thriving SaaS ecosystem in your organization with grit, determination, and a sense of humor. I know because I have done it.

Organization and Chronology

I wrestled with the task of organizing it based on many differing opinions. Some wanted a straightforward, step-by-step journey through the product lifecycle, from product selection to needs interviewing to contract negotiation. While this approach may seem logical at first, it didn't align with my style or reflect the natural, sometimes messy progression of real-life CRM experiences. Questions and issues arise at different times, and case studies come from various stages of the implementation process.

It's important to note that this book is not an implementation guide but a collection of "this just happened to me" stories. The kind that makes you wonder, "I wonder if it happened to others?" So, there is indeed a method to my madness. I advise you to skim the contents table and browse the entire book. See what catches your eye and what resonates with your own experiences, and dive into those sections. Trust me; it's that simple.

Ultimately, I believe that by reading How Not To F*ck Up A CRM, you will be much closer to mastering the art of CRM without making the same colossal mistakes as others. However, luck may not always be on your side, and you may find your own unique way to mess things up. But fear not—I'm here to help. If you find yourself in that predicament, shoot me an email, and let's have a chat. Remember, we're in this together. 😉

Thanks To My Friends and Colleagues

Dear reader,

I'd like to take the time to acknowledge the many individuals who have helped me bring this book to fruition. As someone who has spent decades working on CRM and other large system issues, I knew I had a wealth of knowledge to share with others in my field. The idea for this book had been percolating for some time before I finally put pen to paper in 2022.

Writing this book has been a labor of love, and I couldn't have done it without the support of the many colleagues who have become true friends over the years. Their dedication and expertise have been invaluable to me, and I am deeply grateful for their contributions.

I'd also like to thank the hundreds of technology superstars I've had the privilege of speaking within 2022 and 2023. Their journeys have helped solidify concepts I've experienced firsthand and have added to the pile of valuable case studies I refer to throughout this book.

I want to give special recognition to some individuals who have played a pivotal role in this book. **Susan Tobin** deserves credit for initially recognizing the value of my experience and encouraging me to write this book. **Andrea Ormond** provided crucial support and acted as a daily sounding board, ensuring the ideas I explored were sound. **John Rossman** has been a writing mentor and sanity checker throughout the process, while **Mark Kindy** has helped me stay motivated and provided valuable feedback as a sounding board. Finally, I'd like to express my gratitude to **Linda Orton** and **Andy Pfeffer** for being part of my professional journey. Both **Rusty Atkinson** and **Dave Spencer** are more newly formed professional friends that have given me ideas and support as I venture out on my authorship and expand my CRM and large system recovery services.

In the fast-paced world of being a new-to-be-author, I find solace and gratitude in the unwavering support of my family 🩶. They have been by my side through the ups and downs, enduring my moments of self-doubt and celebrating each milestone achieved.

I cannot express my appreciation to **Karine** for dealing with me and with her unwavering support and putting up with my daily writing saga, **Ashley** for walking into doors and providing comedic relief, **Justin** for strength and learning my secret "code" speak, **Tommy** for picking up on errors that I missed many times over, **Mary** for support, encouragement and nice wishes, **Max** for just being Max, **Elliot** for not asking to be walked too much, **Justice** for her hunger (literal and figurative), and my mom and dad for helping me learn words and sentences and stuff way back when. As I embarked on this journey to write a book, their patience, and unwavering belief in me have been the cornerstone of my persistence. They have provided an anchor of support that has kept me grounded and motivated even in the face of challenges.

Writing a book is no small feat and often requires long hours, late nights, and sacrifices. During these times, family support becomes an invaluable source of strength and inspiration. I am grateful for their understanding, encouragement, and the constant reminder that I can and should continue pursuing my passion.

I am indebted to my family for their unwavering belief in my abilities and continual support through this journey. Without them, this book would not have come to fruition.

As you read through these pages, I hope the knowledge and insights shared within empower you to successfully navigate the challenges and complexities of CRM and extensive system management. I wish that the lessons I've learned - both through triumphs and setbacks - will serve as a guiding light on your professional journey.

Thank you, once again, to my remarkable family for being there every step of the way. Your love and support mean the world to me.

Thank you for reading.

Best regards,

Michael M Hudlow. ☮️,🩶,😁

Sarasota 🇺🇸

Merida 🇲🇽

Cotignac 🇫🇷

About The Author

Michael M. Hudlow

Michael is a Global Conundrum Solver, CRM Champion, and overall international team builder extraordinaire.

As an independent consultant at Bitxia, LLC, Michael works primarily with executive leaders to ensure that their areas of responsibility are always a top priority, allowing them to focus on their areas of expertise without technical concerns or hindrances. Constantly seeking out new challenges that arise when leaders start to doubt the current course of technical actions, using his expertise to either confirm strategies or point out potential areas of failure.

Throughout his career, Michael has honed his skills in team building and operations, management consulting, business development operations, and leading highly matrixed cross-functional teams to achieve successful service delivery and project completion. He is a Salesforce CRM Team Leadership expert, a former licensed title agent in 14 US states, a CBI behavior interviewer and assessor, and a certified SCRUM project owner.

Michael is also an imaginative leader with a big world mindset and a visionary that provides strategic oversight, organizes highly matrixed operations, and develops policies and best practices. He is a trusted advisor and collaborator with key stakeholders, and business units to enhance customer satisfaction and multiply business value while promoting team development, employee engagement, and operational innovation.

Why Now?

Reflecting on my early days assisting companies with CRM installations back in 2009, I realize that I didn't challenge established theories or misconceptions enough. I naively trusted that everything would somehow work itself out without realizing the importance of addressing issues head-on. I failed to insist on lower-level inclusion and more diverse leadership and overlooked that other divisional leaders were not "officially required" to be involved. Yes, I had the green light to move forward through the proper channels, but that was not enough to ensure success.

At the time, the team I worked with was eager for positive outcomes and was keen to provide solutions. I shared their excitement and sincerely desired to see the projects run smoothly and satisfy everyone involved. But now, in hindsight, I realize that my behavior was foolish. Thankfully, I want to equip you with more tools than I had to ensure that your experience in implementing CRM and its usage is the best it can be.

I want to give you a greater chance of success, which means being upfront and explicit about the risks and pitfalls. I want you to have a clear vision of what your company will look like after implementation, not only because it's useful but because I firmly believe that knowing where you're going reduces the chances of getting lost. There are right and wrong ways to approach CRM, and I aim to help steer your organization in the right direction. I want to spare you from the pitfalls that I fell into and provide you with a deeper understanding of what it truly means to be successful in your CRM implementation.

Moreover, if you are embarking on an upgrade or a new software installation, I want to ensure that you are armed with all the necessary information so the process runs smoothly from start to finish. Recognizing that many executives today achieved tremendous success without relying on CRM is essential. Unfortunately, this has led to skepticism about the necessity of such tools and a lack of trust in the power of sharing and transparency for personal and professional growth. While some companies claim remarkable productivity improvements, my experience tells me these claims are not always universally accepted.

In this book, I will draw upon my years of hands-on work and real-life experiences to guide you through CRM implementation's challenges and potential pitfalls. I will provide you with practical advice and proven strategies to help you successfully navigate the complex CRM landscape. By the end, you'll understand the importance of due diligence, realistic expectations, and effective leadership in achieving CRM excellence.

Together, let's transform your CRM implementation journey into a smooth and successful endeavor, helping you unleash your organization's true potential.

The two most significant battles in this space are:

- Persuading C-level executives to **give the system a chance** (which involves more than just talking) and;
- Convincing C-level executives that **they can be successful** in a model where strategy, marketing, innovation, and operational excellence are all part of a single integrated process. Unfortunately, I am unsure if I have ever seen "true" buy-in at this level. By "true buy-in," I mean that behind closed doors and with their teams, they still have many cards held close to their vest.

Leaders who continue to believe, **and more importantly champion**, that technology will help them to be more productive, creative, and profitable are the users of CRM. But many still don't share or understand the value of transparency and collaboration in their success. Many still operate as if the old way of doing things is acceptable. It's not. Leadership today must be more collaborative and inclusive than ever before. This means being open to new ideas and approaches and welcoming those who can help them achieve their goals. So what's the best way to convince these leaders that they need CRM? As a starting point, it's crucial to understand how they see the world right now. We often assume that everyone sees things similarly as we do.

The biggest challenge is getting C-level executives to give the system a chance - for you to convince them to see things the way they could be AND for you to see things through their eyes to best pick the starting point for the story and the ultimate journey.

Note: Working with C-level individuals requires more than just talking:

- C-Level decision-makers **must see it in action and experience how technology can improve their ability to make better decisions**. The theory will not suffice here. Do whatever you can to show examples of other installations. The more hard work that is put in the beginning will pay for itself a thousand times over (maybe a million)

- C-Level decision-makers **need a team of quality business partners** who are familiar with the needs of their organization and understand the style by which they get things done in helping clients solve problems and achieve goals.

The bottom line is that the evolution of CRM is more than just a tool to help people manage their contacts and leads. It's a platform for change that enables users to become more productive, collaborative, and innovative. CRM Technology is not a panacea but a tool that can improve how you run your business. The key is to get everyone to understand how these tools fit into your existing processes and culture.

Unsurprisingly, despite all of the positives, many executives don't believe that CRM solutions are essential, and most think they are way too expensive. C-Level leaders must understand the following points, or there is no way that any implementation or long-term effort will succeed:

- **Without their (C- Level) vocal support, the system will not succeed**. - *There is no compromise on this point*
- **Promoting transparency and sharing ideas** and data (within the operating system) are accelerators for everyone's success.
- Without a modern CRM, **competition will advance quicker** and respond faster to opportunities.
- Eventually, **the company/organization will struggle to keep/attract new talent** without these rapid knowledge tools CRM as younger generations expect it.

CRM has become an essential tool for businesses worldwide, enabling companies to understand their customers better, improve sales, and streamline operations. However, despite its widespread adoption, many misconceptions about CRM persist. In this book, we will explore some of these misunderstandings from various industry perspectives and shed light on the actual value of CRM.

CRM is only for sales teams; While it's true that sales teams often utilize CRMs to track leads and close deals, their scope extends far beyond just sales. Marketing, customer service, and other departments can also benefit from CRM. For example, marketing teams can use CRM data to create targeted campaigns, while customer service representatives can quickly access customer history to provide personalized support.

In the hospitality industry, where customer experiences are paramount, CRMs can help hotel and restaurant staff deliver personalized interactions, track guest preferences, and ensure high-quality service across multiple touchpoints.

CRM is just a fancy pants address book; Though CRMs store contact information, they are far more than digital address books. CRM platforms offer powerful tools for analyzing customer data, identifying patterns and trends, and generating actionable insights.

For instance, in the healthcare industry, CRM platforms can help medical practices track patient records, appointment histories, and billing information. This data can be used to identify trends in patient behavior, such as appointment cancellations or adherence to treatment plans, ultimately enabling healthcare providers to improve patient outcomes.

Implementing CRM is always expensive and time-consuming; While some CRMs can be costly and require considerable time for implementation, this is not always the case. Many affordable and user-friendly options are available, tailored to the needs of small businesses or organizations with limited resources. Cloud-based CRM platforms, for example, have made deploying CRM without extensive infrastructure simpler and faster.

In the nonprofit sector, CRM solutions can help organizations manage donor relationships, track volunteer activities, and streamline fundraising efforts. By choosing the right CRM, nonprofits can reap the benefits of CRM technology without breaking the bank.

CRM is just for large corporations; Though CRMs are

often associated with large businesses, entities of all sizes can benefit from their capabilities. Small businesses, in particular, need to establish strong relationships with their customer base to grow and thrive. CRM platforms can help small businesses understand their customers' needs, preferences, and behaviors, allowing them to deliver personalized experiences that foster brand loyalty. For example, in the retail industry, small boutique shops can leverage CRM to manage customer profiles, track purchase histories, and create tailored marketing campaigns. Such targeted approaches allow these businesses to compete with larger retailers and provide unique customer experiences.

Set Up For Reading This Book

Let's set expectations right from the start. This book will not walk you through the technical nitty-gritty or provide a step-by-step chronology of Customer Relationship Management (CRM) implementations. You won't find detailed instructions on how to set up specific features or configure your system.

Instead, I will delve into the dark corners CRM implementation—the often obscure and less frequently discussed aspects that can potentially throw a massive iron rusty wrench in your plans and create an enormous mess. These are the hidden pitfalls that even experienced professionals can overlook when caught up in the excitement of implementing a new CRM.

Why am I focusing on these less glamorous topics? I've seen first-hand how neglecting these critical considerations can lead to severe headaches and costly mistakes. Trust me when I say these issues can become a significant challenge if not addressed early.

We will explore the potential roadblocks, political hurdles, and unexpected twists that can transform a seemingly smooth CRM implementation into a chaotic mess. By illuminating these less-discussed aspects, we'll help you navigate the potential minefields and give you the wisdom to anticipate and mitigate potential disasters.

Remember, knowledge is power when it comes to CRM implementation. << Yes, it's an overused statement and rather pukey - Sorry. My point is that it's not just about the technical know-how; it's also about recognizing the potential stumbling blocks and taking proactive steps to avoid them. So, grab your coffee, and let's go through the often-overlooked aspects of CRM implementation that could make or break your project.

Nowadays, it is evident that no business can function without a CRM framework, which is a crucial marketing platform. What is different about such a marketing system is that the term "marketing" itself has radically changed in the past few years. My point is that people say some issues are "IT" problems (e.g., email, SMS, printing, word processing, and spreadsheets); well, aren't each of these also marketing? To that point, IT, Marketing, Finance, and HR have become so technically intertwined that they are all morphing into one nerdy glob. So, is CRM marketing, yes. It is also IT. It is also Finance. It is also HR. It is Sales. What the hell isn't it? It isn't any single point and isn't to be taken lightly.

> Bad News - Your customers don't care about your software, and to compound matters, quality CRMs require a significant amount of change management (across all departments) if it has a chance of paying for itself and hopefully actually making opportunities and cash for you).

> Good News -Most issues with getting a CRM off the ground and working as desired are not technical and are frequently not expensive. The main failure points I have uncovered and highlighted in this point can be, at their very core, be considered to be basic project blocking and tackling - BUT - what is different is that it takes much more "soft" skills, point-of-view considerations, understand of the miles walked in each VIPs shoes, convincing of the power of sharing, and true synergies.

As seasoned professionals who have navigated the rocky terrain of CRM implementation, we know all too well the countless ways things can go awry.

The potential pitfalls are numerous and diverse, from botched integrations to poor data quality. But fear not, for this book is here to guide you, esteemed project managers with experience under your belt.

While I acknowledge that this book cannot possibly cover every possible misstep, it does offer invaluable insights into some of the most common mistakes made. Through real-life experiences and lessons learned, I offer mentorship and hard-earned wisdom to help you avoid the same mishaps that have tripped up others before you.

As you dive into these pages, remember that the guidance presented here is geared toward those already familiar with the technical complexities of CRM management. I encourage you to embrace the terminology and concepts presented while ensuring the content remains accessible to non-technical executives.

Above all, I impress upon you the importance of due diligence in avoiding the potential pitfalls laid bare in these pages. Let my experience guide you, and let this book serve as your roadmap to navigating the complex world of CRM implementation and management with ease and confidence.

It's also an excellent resource for new project managers who want to avoid making the same mistakes. I am always looking to improve my knowledge base - so I thought to myself: why not create a book that exposes the errors myself and others (or because of others!)have made and share them (emphasizing the lessons learned)? This book is not about blaming others for their mistakes; it's a way to learn from them. There are many different types of program managers, project managers, champions, sponsors, and users, and this book covers all of them: from newbies to seasoned veterans. It doesn't matter if you manage software development projects or building construction - ensuring the team is on the same page concerning expectations and deliverables is essential. I hope this book is useful and some tips will help you improve your project management skills.

Insider-like Information; With access to a customer's complete history and preferences, small businesses can provide timely and personalized support, improving customer satisfaction and loyalty.

Marketing Campaign Management; Regarding CRM, large organizations have the upper hand with their abundant resources and budget to invest in sophisticated systems. But for small businesses, budget constraints often pose a challenge. However, this hurdle should not discourage small companies from realizing the significance of CRM and seeking out a solution that aligns with their needs. Remember, every step of a small business's journey is pivotal to its ultimate success, and achieving high-caliber results is the expectation. So, small businesses must tread carefully, conduct due diligence, and find an affordable CRM solution that can fuel their growth and enable them to compete with the big players in the market.

Here's why small businesses need a CRM and some practical advice for navigating the challenges of limited resources:

Centralized Customer Data; Like large organizations, small businesses must keep track of customer interactions, sales opportunities, and service requests. With a CRM, small businesses can centralize all customer-related information, including contact details, communication history, and purchase data. This comprehensive view allows small companies to understand their customers better and provide personalized experiences, increasing customer satisfaction and loyalty.

Streamlined Sales Process; A CRM can help small businesses streamline their sales processes and automate repetitive tasks. Small companies can prioritize leads, identify sales trends, and allocate resources effectively using lead management, opportunity tracking, and sales forecasting features. This ensures sales efforts focus on the most promising opportunities and improve overall conversion rates.

Efficient Customer Service; Providing exceptional customer service is crucial for small businesses to build solid relationships and retain customers. A CRM can help manage customer inquiries, track service tickets, and monitor response times. By hitting marketing budgets and need to make every dollar count. A CRM enables small businesses to segment their customer base, target specific groups, and track the effectiveness of marketing campaigns. By analyzing campaign metrics such as click-through rates, conversions, and customer acquisition costs, small businesses can optimize their marketing efforts and achieve better returns on investment.

I can't think of a single organization that couldn't benefit from a more efficient way to manage their contacts, sales processes, intellectual property management, and the myriad of other valuable pieces of information often scattered throughout the company. This crucial data may be hidden in numerous drawers, hard drives, file cabinets, spreadsheets, disorganized notebooks, and countless other obscure locations. The irony is that most people don't even realize they already have a CRM in place, albeit inefficient and disorganized, even if they haven't invested any money into it. Why is this? It's because most CRMs are complicated, expensive, and time-consuming to implement. This has created a large gap between what companies need and what they have to offer.

With a CRM, you can:

- **Manage all of your sales leads** and contacts in one place. Anytime a new Lead comes in or an existing connection needs to be updated, you can access the most up-to-date information at your fingertips.

- **Track opportunities** through each stage of the sales process to ensure that nothing falls through the cracks.
- **Keep track of what offers work for which clients** so that you know exactly what messages resonate with them when you pitch new ones.

It's not uncommon for companies to spend thousands (even millions) of dollars on a CRM only to discover that they already have a crappy one. It is frequently a system held together by manual steps, old Excel spreadsheets, disjointed and horded information and reports made by next-level cut-n-paste methods. The truth is that most businesses are already using many of these as a customer management software/processes and are loosely joined, even if it's not the same one used (or demanded) by their competitors or other successful businesses in their industry.

The problem with this approach is that it's challenging to make sense of the data that has been collected. It's also very time-consuming to look up and retrieve information when needed. This can impede growth and success, especially for small businesses. The problem is that most companies aren't managing their CRM data effectively. They either don't have any system in place, or they use a desktop application to manage it, which doesn't allow them to leverage the power of cloud computing.

Consider this: a CRM won't fundamentally alter the essential information you possess. Instead, it will enhance your awareness of this information and enable you to make faster decisions, ideally outpacing your competition and securing a financial advantage. A well-implemented CRM transforms disorganized data into actionable insights that your current "system" may fail to provide, given its reliance on file cabinets, spreadsheets, and weekly team calls. In other words, your existing "system" likely falls short of delivering the desired results.

By adopting a robust CRM solution, organizations can streamline their data management processes and unlock the full potential of the information they possess. A well-organized CRM centralizes and organizes data, facilitates collaboration among team members, automates routine tasks, and provides valuable insights to drive strategic decision-making. A well-structured and maintained CRM can lead to increased efficiency, improved customer relationships, and a more competitive edge in the market.

Businesses need to recognize the limitations of their current data management practices and understand the value that a comprehensive CRM can bring. By consolidating scattered information into a centralized platform, organizations can eliminate inefficiencies, reduce the risk of data loss or miscommunication, and empower their teams to make more informed decisions. Furthermore, a CRM can help identify new sales opportunities, enhance customer support, and streamline internal communication, all contributing to a more prosperous and profitable business. A comprehensive CRM can help companies achieve their goals by improving how they manage data, process transactions, and engage with customers. Businesses need to recognize the limitations of their current data management practices and understand how a CRM can contribute to their success.

A CRM is essential for businesses of all sizes, but its value is especially apparent for small- and medium-sized enterprises. These companies often lack the resources to hire a dedicated sales team, which makes it difficult to track customer interactions and keep records in sync. A CRM provides a central database where sales representatives can store important information about each sale, including relevant files and documents. It also allows managers to access this data from anywhere, making it easy to monitor what's happening in their organization at any moment.

In most organizations I've seen, the importance of data quality is often understated, overlooked, or downright disrespected. It's like a crown jewel gathering dust in some forgotten corner while everyone's dazzled with the glitz and glamor of new technologies and flashy dashboards. It's time to change that. In the quest for digital transformation, the essence of superior data quality must be understood, respected, and appreciated. Your mission is to foster that culture and spread that understanding. Nowhere in the CRM lifecycle is data quality more vital than when you're setting up a CRM. Here's why: Data in a CRM is not just ones and zeroes. It's your clients' trust. It's your future sales. It's your lifeline to opportunities. It's the magnetic compass that points toward sound decision-making. You mess it up, driving blindfolded on a highway, praying that market trends, customer preferences, or strategic planning don't throw a curveball your way.

"When writing this book, I was like a broken record, constantly typing "customer relationship management," "CRM", "CRMs, "CRM's" and "CRM systems" over and over again. You name it, I typed it! But let's face it, nobody wants to read a book filled with clunky and only minimally different terminology that ultimately means the same thing. So, to keep things simple, I boiled it all down to the trusty three letters - **CRM**!

I know my context may suggest otherwise and that some of you may prefer using it in plural, possessive, or as a comprehensive all-inclusive word to imply everything. But I just used the three simple letters "CRM" as a catch-all. Dont stress out over it, I didn't. Maybe you call it a system, process, or workflow? Whatever floats your boat! It is essential to understand and use it appropriately in your own situation. I think we all know what are all talking about.

Sensitivity and Language Usage

In the world of CRM, we often find ourselves navigating complex challenges and encountering frustrating situations. Through my years of experience, I can deeply sympathize with CRM professionals' struggles at all levels. Addressing the sensitivity surrounding language usage in this book is essential.

To some readers, it may be apparent that within the "CRM Club," a unique camaraderie and understanding exist. I have had the privilege of sitting in countless rooms with CRM professionals, witnessing the dedication, hard work, and perseverance they bring to their roles. As a community, we are all in this together, and I want to convey this sense of togetherness in these pages.

Now, speaking frankly, I must admit that I occasionally employ colorful language in this book. However, it is essential to understand the context behind it. When I use words that may raise eyebrows or call out certain behaviors, it is done so in a spirit of fun and as a term of endearment within our unique professional environment. These words are not meant to be taken literally or as personal attacks on individuals but rather as a way to collectively address shared frustrations and hurdles.

In reality, those we encounter in our CRM journey are our colleagues and partners in achieving success. However, during the heat of corporate project battles, it is not uncommon for tensions to run high. In those challenging moments, I might refer to specific individuals as, let's say, "challenging characters." Please understand that this is to convey the intensity of our struggles and does not reflect my actual personal opinion or characterization of those involved.

By acknowledging the sensitive nature of language usage and clarifying its purpose within the context of this book, I hope to strike a balance between authenticity, camaraderie, and the professionalism required within the CRM field. Together, we can navigate the challenges, learn from one another, and ultimately succeed in our CRM endeavors.

As professionals and as a community, let's proceed on this journey to uncover insights, share experiences, and make the most of our CRM efforts.

Part I: Evaluating and Pre-Work

In the first section of this book, I'll delve into a crucial aspect of CRM implementation - the evaluation process and the associated topics that come into the picture during this phase. Whether considering a completely new CRM or assessing a significant add-on for your core system, this section will cover topics requiring severe consideration. It's important to note that these topics can disrupt your plans and lead to unexpected stress if not adequately addressed.

During the evaluation phase, numerous challenges can arise that can turn things upside down. It's essential to approach this process with caution and diligence. This section will help you navigate the complexities and make informed decisions, ensuring you're prepared for each step's potential impact and consequences.

It's worth emphasizing that the topics discussed in this section (and the others sections as well) are not limited to the a single phase alone. They can surface at any point in the CRM lifecycle, from implementation to ongoing maintenance and upgrades. By gaining a deep understanding of these topics early on, you'll be better equipped to handle unexpected issues that may arise down the line.

Remember, the goal here is to empower you with the knowledge and insights to make informed decisions and anticipate potential challenges. Pondering these topics and taking them seriously will help you avoid unnecessary stress and proactively address any roadblocks that may come your way. So, buckle up, grab your favorite mug of coffee, and dive deep into the world of CRM evaluation and its long-lasting impact.

1

Your F*cked If VIPs Aren't Really "Sold" On The Effort

The Role of Management Support in CRM Success

BEFORE LAUNCHING A CRM, it is essential to secure strong leadership support to ensure its success. Regardless of their desire, leadership plays an enormously vital role in ensuring the all-encompassing adoption and effectiveness of CRM within any organization. You can create a positive environment that fosters collaboration, innovation, and growth by actively engaging leaders and turning them into vocal champions of the CRM.

WARNING: Passive, not active, leadership engagement is an early warning sign of lack of buy-in and/or "true" commitment to the effort. This happens a lot and should be fleshed out in the beginning of any CRM effort and not when you need their support on an important item in the future.

BELOW ARE strategies to solidify leadership support for the CRM and ways to encourage them to become active and vocal champions:

Demonstrate The Strategic Value Of The CRM; Show leaders how the CRM aligns with the organization's strategic goals and objectives and can help drive growth, improve efficiency, and enhance customer relationships. Collect and provide case studies, research, and success stories from other organizations successfully implementing CRMs. Provide leaders with regular updates on the CRM's progress, highlighting key milestones, achievements, and their positive impact on the organization.

Involve Leaders In The Planning And Implementation Process; Engage leaders in decision-making, giving them opportunities to contribute their ideas, insights, and expertise. Assign leaders specific roles and responsibilities related to the CRM's implementation, ensuring they are vested in its success. Encourage leaders to participate in training sessions, workshops, and other CRM-related activities, helping them develop a deep understanding of the system and its capabilities.

Communicate The Benefits Of The CRM To Leaders And Their Teams; Clearly articulate the advantages of the CRM for leaders and their departments, focusing on the features and capabilities most relevant to their needs and goals. Offer personalized demonstrations and training sessions to help leaders and their teams understand how the CRM can improve their day-to-day work and drive better results. Share success stories and testimonials from team members who have embraced the CRM and experienced positive outcomes.

Foster A Culture Of Collaboration And Innovation; Encourage leaders to promote a culture of collaboration, open communication, and shared responsibility for the CRM's success within their teams. Provide leaders with the tools, resources, and support they need to drive innovation and continuous improvement within the CRM. Recognize and reward leaders who actively champion the CRM and demonstrate a commitment to its success.

Establish Clear Expectations And Accountability; Set clear expectations for leaders regarding their involvement in the CRM's implementation and ongoing management, including specific goals, deadlines, and performance metrics. Hold leaders accountable for meeting these expectations, providing regular feedback and support to help them stay on track and address any challenges. Celebrate the achievements and progress of leaders who actively support the CRM, reinforcing the value of their efforts and encouraging continued engagement.

A LACK of public support from leadership can significantly negatively affect the CRM's progress and overall success. Below are five ways that a lack of leadership support can harm the CRM's progress and the cons of doing nothing to change this:

Reduced Adoption And Engagement; Without solid leadership support, team members may be less likely to adopt and engage with the CRM, potentially limiting its effectiveness and benefits. If the lack of leadership support is not addressed, the CRM may fail to reach its full potential, resulting in wasted resources, reduced efficiency, and missed opportunities for growth.

Resistance To Change; When leaders do not actively support the CRM, they may inadvertently contribute to a culture of resistance to change, making it more challenging to implement new processes and technologies. If this resistance is not addressed, the organization may struggle to adapt and innovate, ultimately hindering its ability to compete and succeed in the market.

Decreased Morale And Motivation; A lack of visible leadership support can decrease morale and motivation among team members, who may feel unsupported and undervalued in their efforts to adopt and utilize the CRM. If morale and motivation continue to decline, the organization may experience increased turnover, reduced productivity, and a decline in overall performance.

Inconsistent Implementation And Usage; Without clear expectations and accountability from leadership, the implementation and usage of the CRM may be inconsistent across the organization, leading to gaps in data, reduced efficiency, and a lack of collaboration. If these inconsistencies are not addressed, the CRM may fail to deliver the expected benefits, resulting in wasted resources, missed opportunities, and decreased customer satisfaction.

Limited Organizational Buy-In; An increased challenge will arise if leaders do not actively champion the CRM, securing buy-in from other organizational stakeholders, including employees, customers, and partners. If the organizational buy-in is not achieved, the CRM may struggle to gain traction and perform its intended goals, ultimately limiting its impact on the organization's success.

BY PROACTIVELY ADDRESSING the potential challenges posed by a lack of leadership support, organizations can ensure a smoother CRM implementation process and maximize its benefits. This will improve efficiency, improve customer relationships, and increase overall success.

As discussed, management buy-in and support are paramount in successfully implementing and using CRM. The problems may not be noticed immediately, but the cracks will show in time, maybe in a few months or years.

Successful CRM implementation and ongoing use depend on various factors, one of the most critical being management buy-in and support. Without adequate management support, organizations may face numerous challenges that can hinder the effectiveness of their CRMs and, ultimately, impact their bottom line. Their buy-in refers to top management's commitment and active involvement in the decision-making process, implementation, and ongoing support of a CRM.

This support is crucial for several "Ongoing Reasons":

Resource Allocation; Implementing a CRM requires significant time, money, and resources. Management buy-in ensures that the necessary resources are allocated to the project, including budget, personnel, and technical support. Without adequate resources, the implementation process may be delayed, incomplete, or unsuccessful, leading to wasted investments and lost opportunities.

Organizational Alignment; A CRM impacts various departments, from sales and marketing to customer service and IT. Management support is crucial for aligning these teams and ensuring everyone is on the same page regarding the CRM's goals, objectives, and expected outcomes. This alignment helps to prevent potential conflicts, miscommunications, and inefficiencies that can arise when different departments have competing priorities or are not fully committed to the CRM.

Change Management; CRM implementation often requires significant changes to existing processes, workflows, and organizational culture. Management buy-in is essential for driving these changes and ensuring that employees understand the reasons behind them and the benefits they can expect from the new system. Without management support, employees may resist the changes, leading to a lack of adoption and, ultimately, a failed CRM implementation.

Employee Training And Adoption; For a CRM to be successful, employees must be trained to use it effectively and encouraged to adopt it daily. Management support is crucial for providing the necessary training resources and fostering a culture of continuous learning and improvement. Additionally, management can serve as role models by actively using the CRM and demonstrating its value to the organization.

Ongoing Support And Optimization; The work does not end once a CRM is implemented. Continuing support and optimization are necessary to ensure the system meets the organization's needs and delivers the desired results. Management buy-in is essential for providing the resources and commitment needed for ongoing system maintenance, updates, and improvements.

. . .

WHEN MANAGEMENT SUPPORT IS INADEQUATE, organizations may encounter several problems that can impede the success of their CRMs:

Insufficient Resources; Without management buy-in, the CRM implementation may suffer from a lack of resources, leading to delays, inefficiencies, and potential system failure.

Lack Of Organizational Alignment; Inadequate management support can result in a lack of alignment between departments, leading to conflicts, miscommunications, and inefficiencies that can hinder the effectiveness of the CRM.

Resistance To Change; Without management support, employees may resist the changes required for CRM implementation, leading to a lack of adoption and, ultimately, a failed system.

Poor Employee Training And Adoption; Inadequate management support can result in insufficient training resources and a lack of encouragement for employees to adopt the CRM, leading to suboptimal usage and reduced benefits.

Inability to Optimize And Maintain The System; Without management buy-in, organizations may struggle to allocate the necessary resources for ongoing system maintenance, updates, and improvements, leading to decreased system performance and effectiveness over time.

MOVING PAST NAYSAYERS, business hurdles, and financial concerns by division leaders requires unwavering management buy-in and support are crucial factors in the successful implementation and ongoing use of CRMs. By providing the necessary resources, fostering organizational alignment, driving change management, and promoting employee training and adoption, management can help ensure that their CRM delivers the desired results and drives business growth. Organizations that fail to secure adequate management support may face numerous challenges that can hinder the effectiveness of their CRMs and negatively impact their bottom line.

Part I: Evaluating and Pre-Work

2

Case Study: Death By Committee - I/We Are Special

Scene:

As seasoned professionals in CRM, I recall countless scenes of the hustle and bustle of a fast pasted office, the palpable energy of a team united in pursuit of a singular goal, the excitement of collaboration, and the eagerness to contribute one's expertise and ideas – all familiar elements in the challenging journey of designing a CRM that can accommodate the diverse needs of various stakeholders.

Yet, I caution those in this scene or a similar one that the journey ahead is filled with challenges that demand careful consideration and due diligence. Collaboration is undoubtedly a buzzword in this setting. Still, it should not come at the cost of a thoughtful, data-driven approach that aligns with your ultimate goals and priorities.

Incident:

Amidst the once-promising atmosphere of the project, chaos began to permeate as the third-party system implementer entered the scene. Assigned with facilitating coordination and communication among the team, this individual's approach unknowingly unleashed a storm of disorder. As the meetings ensued, it became apparent that the implementer had misguided intentions or was feeling overwhelmed and resorted to throwing questions to the entire group without considering staff level or role. The consequences were swift and disheartening – haphazard and often uninformed responses dominated the discussions.

The implementer's indiscriminate questioning turned the atmosphere into a scramble in what were meant to be productive and focused meetings. Individuals were now offering opinions and suggestions on matters they had yet to gain the necessary experience or authority to address and/or may not even be relevant to the mission at hand. As a result, egos collided, and the project quickly became a battle of conflicting viewpoints. Pride took center stage, overshadowing the pursuit of practical solutions that could propel the project forward.

Unbeknownst to all, this incident was proof of the importance of thoughtful leadership and effective communication in such projects. The role of a third-party system implementer is not merely to manage logistics but also to foster an environment of collaboration and expertise. Their ability to guide discussions, consider the levels of knowledge and experience within the team, and facilitate focused conversations is of utmost significance.

Result:

Despite the initial promise of such endeavors, all watched them unravel into a tangle of complexity, directionless and ineffective, and a shell of the intended purpose.

Regrettably, misguided meetings resulted in a subpar system failing to deliver on its intended goals. Instead of a product that satisfied everyone's needs, a disjointed and confusing system in desperate need of further development emerged. An unchecked exchange of opinions and a disregard for expertise and rational decision-making profoundly influenced the path chosen, eroding the effectiveness and impact the system could have had.

Lessons Learned:

I cannot emphasize enough the importance of due diligence, expertise, and disciplined decision-making in every phase of the CRM journey. As professionals, we must prioritize collaboration, informed decision-making, and an environment where ideas are exchanged openly and scientifically vetted.

Purposeful collaboration is critical; While collaboration is essential, it is crucial to approach it with a clear sense of purpose. Delineate specific roles and responsibilities within the team, ensuring that individuals with the necessary expertise make decisions about their respective domains. Respectfully validate and evaluate opinions to foster focused collaboration rather than allowing unqualified voices to hinder progress.

Expertise should guide decisions; Recognizing and honoring the expertise of individuals within the team is vital for success. Encourage an open dialogue where opinions are only given weight if grounded in knowledge and experience. Foster an environment where individuals feel comfortable deferring to others when they lack the necessary expertise.

Quality over quantity in meetings; Limit the number of attendees, ensuring that only those directly involved in the specific topic or decision are present. Smaller, focused meetings allow for more productive and efficient discussions, minimizing the chaos and confusion arising from a large gathering of voices.

Embrace humility and open-mindedness; Pride can be detrimental to decision-making processes. Encourage a culture of humility, where individuals are willing to acknowledge when they don't have the answer and are open to learning from others. Seek diverse perspectives and prioritize the collective goal of creating the best solution rather than personal agendas.

Effective facilitation is crucial; The role of a facilitator or system implementer in meetings cannot be underestimated. It's vital to have someone who can guide discussions, structure questions effectively, and manage group dynamics to ensure productive outcomes. Invest time in finding a facilitator who understands the project's scope, is skilled at managing diverse opinions, and can maintain focus and clarity throughout the process.

What should have been done differently?:

Having a clear sense of purpose when collaborating with team members to accomplish a common goal is critical. Unfortunately, in some instances where collaboration is encouraged, chaos ensues, leading to a lack of direction, unchecked opinions, and unproductive outcomes.

To avoid such adverse situations, it's essential to take proactive steps that guarantee project success. If you are in a situation where you must coordinate a team's efforts, we have some practical advice for you.

To begin with, ensure that everyone is aware of their specific roles and responsibilities in the project. This will guarantee that those with expertise make decisions about their respective domains, leading to a more productive and successful outcome.

Further, since expertise should guide decisions, encourage an open dialogue where opinions are only given weight if grounded in knowledge and experience to foster an environment where individuals feel comfortable deferring to others when they lack the necessary expertise.

- **Limiting the number of meeting attendees** is essential to ensure that only those directly involved in the specific topic or decision are present and to prevent unnecessary distractions and delays.

- **Smaller, focused meetings** allow for more productive and efficient discussions and minimize the chaos and confusion arising from a large gathering of voices, resulting in more precise decisions and actionable outcomes.
- **Embrace humility and open-mindedness**. Encourage a culture of humility, promoting a willingness to acknowledge when individuals don't have answers and open-mindedness for learning from others, as this encourages collaboration and innovation.
- **Always seek diverse perspectives** to create the best solution and prioritize the collective goal rather than personal agendas, as diverse viewpoints lead to more robust and well-rounded solutions.
- **Invest time in finding the right facilitator** who understands the project's scope, is skilled at managing diverse opinions, and can maintain focus and clarity throughout the process, as their expertise and guidance are crucial to running effective and efficient meetings.

As you work on your project, keep these proactive steps in mind to prevent unnecessary harm to your project. Applying these lessons allows you to navigate project management challenges more effectively, fostering productivity, innovation, and success.

3

Honestly, Honest

The honesty of your internal customers is vital, and many, unfortunately, are not honest as they initially, or even on an ongoing basis, do not see the need to assist. You must figure out a way(s) to identify them and flip their attitude and opinion.

One of our primary responsibilities is to ensure the success of any significant system installation and operation. In this case, I focus on a CRM project by encouraging project sponsors to be more honest with their concerns and needs. This transparency is crucial to addressing potential issues early on and ensuring the project meets the organization's objectives. Here, I will outline several strategies to foster open communication and engagement from project sponsors.

Address Concerns And Needs Proactively; When project sponsors share their concerns and conditions, addressing them proactively and demonstrating that their input is valued is essential. This may involve conducting further research, seeking expert advice, or implementing changes to the project plan. By taking action to address concerns and needs, we can show project sponsors that their honesty is appreciated and encourage them to continue being forthcoming with their problems and needs.

Develop A Culture Of Trust And Openness; Building trust and transparency within the organization is essential for encouraging project sponsors to be honest about their concerns and needs. This can be achieved by promoting transparency, fostering open communication, and creating an environment where people feel comfortable sharing their thoughts and opinions. As a CIO, I will lead by example, demonstrating honesty and openness in my interactions and encouraging others to do the same.

Establishing Clear Communication Channels; Effective communication is critical to the success of any project. Clear communication channels ensure project sponsors can easily share their concerns and needs with the project team. This may include setting up regular meetings, creating a dedicated communication platform, or implementing an open-door policy. By providing multiple avenues for communication, we can encourage project sponsors to be more forthcoming with their concerns and needs.

Encourage Active Participation; Project sponsors are likelier to be honest about their concerns and needs if they feel actively involved in the CRM project. This can be achieved by assigning specific roles and responsibilities to sponsors, inviting them to participate in decision-making processes, and providing regular updates on project progress. By involving project sponsors in the project's day-to-day activities, we can foster a sense of ownership and responsibility, which will, in turn, encourage them to be more honest about their concerns and needs.

Provide Education And Training; A lack of understanding about the CRM project may make project sponsors hesitant to share their concerns and needs. Education and training sessions can help bridge this knowledge gap and empower project sponsors to make informed decisions. This may include workshops, webinars, or one-on-one coaching sessions to ensure project sponsors thoroughly understand the CRM, its benefits, and potential challenges.

Foster Collaboration And Teamwork; Encouraging cooperation and teamwork among project sponsors and the project team can help to create an environment where honesty and openness are valued. This can be achieved by organizing team-building activities, promoting cross-functional collaboration, and encouraging open dialogue between project sponsors and the project team. By fostering strong working relationships, we can create a supportive environment where project sponsors feel comfortable sharing their concerns and needs.

Implement Feedback Loops; Creating a feedback loop allows project sponsors to share their concerns and needs regularly and ensures that their input is incorporated into the project plan. This may involve setting up a dedicated feedback channel, conducting regular surveys, or organizing focus groups. By implementing a feedback loop, we can continuously improve the CRM project by incorporating the concerns and needs of project sponsors.

Recognize And Reward Honesty; Recognizing and rewarding project sponsors for their honesty can help to encourage them to be more forthcoming with their concerns and needs. This may involve acknowledging their contributions during team meetings, providing public recognition, or offering incentives for honest feedback. By recognizing and rewarding honesty, we can create a positive reinforcement loop that encourages project sponsors to continue sharing their concerns and needs.

Be Responsive And Adaptable; Being responsive and flexible to the concerns and needs of project sponsors is crucial to maintaining their trust and encouraging honesty. This may involve changing the project plan, adjusting timelines, or reallocating resources as needed. By demonstrating a willingness to adapt to the concerns and needs of project sponsors, we can show that their input is valued and encourage them to continue being honest about their problems and needs.

Lead By Example; It is essential to lead by example and demonstrate the importance of honesty and openness in the CRM project. This may involve openly sharing my concerns and needs, actively seeking feedback from project sponsors, and being responsive to their input. By leading by example, I can set the tone for the organization and encourage project sponsors to be more honest with their concerns and needs.

. . .

FOSTERING honesty and openness among project sponsors is crucial to the success of our CRM project. By implementing these strategies, we can create an environment where project sponsors feel comfortable sharing their concerns and needs, ultimately leading to a more successful project outcome.

4

Case Study: People Not Being Completely Honest

Scene:

You're an experienced project manager for a large organization, having seen several successful and failed projects over the years. You've learned the hard way that hidden agendas can harm a project's success and the importance of ensuring aligned hopes and expectations among all team members.

Incident:

As the project team begins to work on a new project, you identify several liaisons who quietly appear to have their own motives and agendas. These liaisons often try to bolster their position within their division or geography, acting like know-it-alls and making double promises. The outcome of these false promises becomes evident once their declarations do not align with those of the core team. This misalignment could ultimately lead to modifications or revisions to the initial plan that satisfies the division leadership's expectations rather than what was initially agreed upon.

Result:

The presence of hidden agendas within the liaisons ultimately creates tension within the project, resulting in the project team's members being forced into decisions that could compromise the project's success. In some cases, this could lead to resentment and disinterest from other parties not privy to the situation.

Additionally, the presence of internal administrative team representatives that behave similarly to Liaisons could present a layer of hidden complexity where they leverage blame on external systems and processes rather than addressing underlying organizational shortfalls and inefficiencies.

Lessons Learned:

Be Alert; Be ready to identify hidden agendas in your project's liaisons. Look out for members who act independently of the team's goals and those attempting to boost their status within the organization.

Ensure Clear Communication; Establish clear expectations and communicate frequently amongst the team and its members. Ensure that the project team leadership, system Administrators, division heads, and internal administrative team representatives clearly understand each task and its expected outcome.

Assure Alignment; Ensure everyone is aligned towards the same goals and resist any tendencies to overpromise or under-deliver.

What Should Have Been Done Differently?

To avoid project setbacks related to hidden agendas, it is essential to:

Meet With Division Leaders Yourself; Establish an open

line of communication. If you suspect something is "odd," call the divisional leader yourself, unannounced, and have an impromptu chat. Dig a little deeper to flesh out any peculiarities.

Establish Clear Guidelines And Procedures; Establishing clear guidelines and procedures creates a level playing field for all team members of projects. This process ensures that every part of a project is documented, discussed, presented transparently, and agreed upon, minimizing the risk of hidden agendas.

Work To Identify Fault Lines: Encourage the people involved in projects to be aware of hidden agendas and to work promptly to identify and address them. When fault lines are identified early, the organization can correct each missed target, causing teams to work towards meeting each target with an understanding of the project's overall health.

Prioritize Open Communication: Maintain open and honest communication among team members. Promote feedback, individual goals, objectives, and suggestions from all parties to ensure that concerns are addressed, and solutions are developed collaboratively.

HIDDEN agendas within teams can challenge any project's long-term goals enormously. The organization can create a quality control system at the onset to mitigate this risk and maintain transparency for all aspects of the working project. Encouraging a culture of openness and honesty and establishing clear guidelines and procedures will minimize risks and promote accountability across all parties.

5

Case Study: Empty Invites

Scene:

I was invited to a regional meeting for the US West Coast to give a large group of leaders an intro to what they could do with Salesforce based on what was developed and actively available to them. I highlighted significant changes that saved large amounts of administrative effort and time. Additionally, the new system was easier to use and provided timely information on their PCs and mobile devices. It was a no-brainer as nothing had to be created especially for them; it was all done. They had to stop using their old Excel spreadsheets that used the "Sneaker Network" (my term for running/sending files and information around using any means possible) to get around and start using modern CRMs. Yes, it looked different, but it was incredibly easier. Before I set out from Florida to San Francisco, I was provided my plane tickets, transfers, and hotel stay by my host, the global leader of the team. He supported the effort. He was spending significant money on the system itself, all of this "proof" allowed me to speak during his team's precious meeting time and covered all of my significant development and travel expenses. He is a brilliant and nice man, and I knew I could count on him to help me help his team to hit a home run. *F*ck, was I wrong.*

The Incident

I felt confident and well-prepared as I prepared for my presentation, getting mic'd up and reviewing my notes. I had done my research, understood their project challenges, and knew their history with outdated systems. Moreover, I had the support of their leaders, making me believe that this presentation would be a breeze. However, things took an unexpected turn as the previous speaker left the stage, and it was my turn to shine.

The emcee introduced me, and as I took the stage, the audience greeted me warmly. I launched my presentation on the *jumbotron*, excited to share the benefits and possibilities of the new system. However, it was at that very moment when everything went awry. The leader who had invited me and sponsored my presence stood up from the front row, accompanied by his first-in-command, and visibly walked out of the room. The room fell silent as everyone witnessed this unfortunate incident. It was clear that my efforts had been undermined in a big way.

Nonetheless, I continued with my presentation, trying to maintain composure. The audience engagement was high, with many questions and murmurs of excitement about the potential time savings and ease of accessing information. However, deep down, I knew this incident had a far-reaching impact. It unveiled a sobering truth - the glimpse of the future they had witnessed clashed with their leader's expectations.

Through his actions, their leader implicitly communicated that while he accepted the concept of progress, he remained firmly rooted in the past. Despite the new system's potential, it became evident that he still expected to use Excel spreadsheets. While he promised not to hinder progress and acknowledged the value of this remarkable tool, his actions spoke louder than words. Behind the curtain, he seemed indifferent and unconcerned about the new system's success. This unfortunate incident effectively condemned people to operate in two parallel systems, forcing them to adopt the new one while still adhering to their preferred old reporting methods.

It was disheartening as I witnessed the clash between a promising future and resistance to outdated practices. However, I remained determined to advocate for progress, believing that the potential benefits of the new system could eventually sway even the most stubborn of leaders.

*What a d*ck* 🦆

What's The Big Deal?

Allies are frequently thought of as friends, supporters of your efforts for the collective good of the users and operators. They should be trusted to support your effort(s) when you are not in the room to explain and/or defend yourself. More importantly, they should be people to openly promote the benefits of the change you and your teams are leading and debunking false thoughts that may pop out of the woodwork when you are not there. Lastly, and most importantly, if these allies are senior and respected, they should be counted on to open doors **(and keep them open)** so that the project team can focus on the task required to get things done. Fortunately, you have been reading this book long enough to know I am about to lob a turd bomb over the fence. Get out your umbrella.

Allies in the business world are not much different than many allies in the military/political world. They must always be massaged. And if you don't maintain them, satisfy them, and ensure you understand and address their whims, then assume they are no different than anyone else. Maybe worse, especially when considering project work requiring timely deployments, change of process, and training. You can not let your guard down or reduce your attentiveness to an ally after you have their support. If anything, you need to pay more attention to them. Your attention can be focused on ensuring their happiness and instructing them on where to help you best, but never forget that they expect something from you. Always.

Outcome

FAIL 😱- Let's take a closer look at this situation. This effort had all the makings of a success story, with a well-executed presentation, supportive materials, and even an invite from the global leader.

However, despite these initial advantages, it never got off the ground. In this scenario, at least two crucial missteps caused things to falter:

- The global leader failed to sit in the room with their staff, witness what their team saw, and hear what their team asked. Without this firsthand experience, it was near impossible for the leader to devise a plan that met their team's needs and aligned with their goals.
- Even though the leader saw the potential for dashboards and reports from new tools, this vision became lost regarding actual implementation. Ultimately, it was too easy for the leader to continue doing things the way they always had, with their staff's happiness sadly not valued enough.

From the staff's perspective, they were excited about the new tools and eager to use them. However, why would they change without a mandate to fully integrate these tools and the looming thought of continuing to use the old system? Ultimately, their hands are tied and accountable to their leader's direction and requirements. When efficiency is jeopardized by the need to use old and new tools, there is no incentive to adopt new technologies.

What appeared to be a slam dunk ultimately gave way to an absolute strikeout. The key takeaway here is that even with the most promising of plans, execution, and follow-through are paramount. Foresight, active listening, and commitment to change are required to ensure that the potential benefits of new technology are not lost during implementation. By staying true to these tenets, organizations can stay ahead of the curve, meet their user's needs, and realize the full potential of their CRMs.

Lesson Learned: Guard Against Complacency

In reflecting on the situation, I realized that I fell victim to the dangerous trap of complacency. The overwhelming success of the event and the active participation had blindsided me, causing me to overlook the need for a healthy dose of skepticism about the "real" next steps. Instead of rushing back to the airport, assuming everything was in order, I should have stayed in town and made the extra effort to ensure that all the hard work had not been in vain.

It's crucial to remember that the pursuit of success is a continuous journey. We should never become complacent, even in the face of apparent triumph. Success is not written in stone, and its meaning can often be distorted or misunderstood. We must remain vigilant, constantly challenging ourselves and questioning the reality of what we perceive as a win.

What I Could Have Done Differently: Building a Safety Net

Suppose I had the benefit of a crystal ball. In that case, there are several actions I would have taken to avoid the pitfalls of complacency:

Coordinated Expectations; I would have proactively coordinated with the event's leader, ensuring they joined me on stage after my talk to set clear expectations for the next steps. This partnership would have helped align our understanding of what success meant and avoided potential misunderstandings.

Expressing Concerns; I should have contacted the leader immediately after the session and candidly expressed my concerns. It's important to communicate openly and honestly, especially when doubts have already crept into our minds. By voicing my apprehension, we could have worked together to address potential issues or gaps, ensuring that the necessary actions were taken promptly.

Developing a Remediation Plan; Once we had acknowledged the concerns, creating a plan to remedy the situation would have been crucial. This plan would have involved identifying potential gaps or improvement areas and taking the necessary steps to rectify them. By embracing a proactive approach, we could have mitigated possible negative consequences and ensured a seamless path to success.

THE LESSONS LEARNED from this experience reinforce the need to **remain vigilant, question perceived wins, and challenge our assumptions.** Pursuing success never ceases, and we are responsible for safeguarding against complacency. By coordinating expectations, expressing concerns, and developing remediation plans, we can confidently navigate the uncertain waters of success and mitigate potential pitfalls.

6

Customers

WHAT'S THE SOLUTION?

The solution is to establish a solid foundation for your CRM project. The customer may always be correct, but the real question is *who your customer is concerning the CRMs usage, interaction, and expectation of improvement.* More than likely, you have dozens, if not hundreds, of customers. From the perspective of a person in charge, or in some capacity responsible, for the CRM, you have to consider every internal department, every industry team, every product team, every salesperson or consultant, every geography, and so on as your customer. Of course, the company's external customers are also your customers.

YOU ARE **the only one who can define who your customer is.** They say you can't please all of the people all of the time. Perhaps that's because many companies don't even know their actual customers or who their customers could be. It could be your boss, your colleagues, your family members, or other people in general. That's why it's essential to define precisely who your real customers are and what they want from you as a company. Successfully defining your ideal customer means understanding what being one means and setting expectations so everyone in the company is on the same page about this definition. Therefore, everyone can work together toward a common goal: delivering outstanding customer value. Make sure that your purpose of each customer's expectations is clear, accurate, complete, **and agreed upon** - Once your real customers are identified, you must ensure that what they want from you as a technology provider (if internal, they are an internal customer) or what is expected from your company (if external), and how well your products and services are currently serving them. It's important to note that these steps aren't meant to be taken in any priority or order; they should all be done continuously and simultaneously.

The best way to do this is by creating a Who are my "real" customers plan and what will satisfy them. Such a plan may seem obvious, and you may believe that you already know this information, but this is chronically and tragically not the case. If you partake in a journey to seek out your "true" customers, you can find the squeaky wheel you never knew about.

Who and How Many Customers Does This Effort(s) Have?

When considering the customers impacted by your efforts, it's crucial to understand who they are and how many different customer types exist.

Let's break it down:

Who are your customers?

Your customers can be both internal and external stakeholders. Internal customers are individuals or groups within your organization who rely on the outcomes of your efforts. External customers are those who directly interact with your products or services.

How many customer types do we have?

It's essential to recognize the different customer types within your organization to tailor your efforts accordingly. Here are some key customer types:

INTERNAL CUSTOMERS:

Divisions; Each division within your organization can be considered its customer type. This could include departments like Sales, Operations, or Customer Service. It's crucial to treat each division separately and obtain buy-in from their key stakeholders. This ensures that your efforts align with their specific needs.

Geographies/Regions; If your organization operates in multiple locations or serves various regions, you should treat each geography or region as a distinct customer type. They may have unique requirements, regulations, or market conditions. Engaging with key stakeholders in each geography or region is vital to ensure their participation and support.

Internal Operating Groups; Departments such as Finance,

IT, Marketing, HR, and others represent separate customer types within your organization. These groups have distinct needs and objectives. When planning your efforts, be mindful of treating each internal operating group as a separate customer type. Secure their buy-in and involve them to avoid stepping on any toes.

VIP/Power Users/Sticks-In-The-Mud/Assholes; These individuals significantly influence your organization. They could be highly valued customers, influential decision-makers, or individuals who raise concerns and challenge the status quo. Getting them on your side is crucial to addressing their needs and concerns. Building positive relationships and involving them in your efforts can help drive adoption and create advocates for your initiatives.

EXTERNAL CUSTOMERS:

Customer Wants And Needs; Understanding the desires and needs of your external customers is crucial. Conduct market research and gather feedback to identify their pain points, preferences, and expectations. This knowledge will help you align your products or services to meet their requirements better.

Interactions With Each Other; Consider how your external customers interact. Are there collaboration opportunities or dependencies among them? Understanding these dynamics can help you design solutions that foster collaboration and enhance the overall customer experience.

Efficiency And Benefits; Lastly, assess how your external customers can benefit from using your products or services more efficiently. Identify areas where your offerings can streamline processes, save time, or provide additional value. By showcasing these benefits, you can increase customer satisfaction and loyalty.

UNDERSTANDING your various customer types and their unique needs is fundamental to success. Treat each category separately, obtain their buy-in, and tailor your efforts to address their requirements. You can build strong relationships and drive positive outcomes for your organization by engaging and delivering value to internal and external customers.

When implementing or optimizing a CRM, it is crucial to identify everyone who should be considered a customer of the system, both now and in the future. Your ultimate goal is to ensure the success of the CRM by catering to all stakeholders' needs. To achieve this, active listening and genuine empathy must take center stage.

Active listening means genuinely hearing and understanding the concerns and needs of your internal customers. Creating a safe space where they feel heard and valued is essential. Neglecting to listen to any person or group can become a chronic issue that undermines the success of the CRM. It is only fair that imposing or selling a system that fails to meet their needs would lead to upset and resistance.

To mitigate this, fostering an environment of open communication and collaboration is crucial. Actively seek feedback and insights from your internal customers, taking their suggestions and concerns to heart. By incorporating their perspectives into the system's design and functionality, you can ensure it becomes a valuable tool that aligns with their requirements and expectations.

Maintaining authentic empathy throughout the process is equally important. Put yourself in the shoes of your internal colleagues and try to understand their pain points and challenges. Acknowledging and addressing these concerns can build trust and rapport, fostering a positive reception of the CRM.

Remember, the success of a CRM relies on more than just its technical features; it hinges on the level of engagement and support from the people who will be using it. By prioritizing active listening and genuine empathy, you can create a culture where the CRM thrives, driving productivity and delivering tangible value to the organization.

7

Ideas On Customers (Users)

When it comes to identifying your internal customers, there are several strategies you can employ. Let's break them down and discuss the pros and cons of each.

First, you can conduct an **organizational chart analysis**. This involves reviewing the organizational chart to identify the departments and teams that rely on your IT services. The pros of this approach include the ease of identifying key stakeholders. However, one caveat is that it may not capture informal relationships or ad hoc requests, which are common in many organizations.

Another method is to **review your IT service catalog**. By analyzing the catalog, you can determine which services different departments use. This provides a clear picture of service usage. However, it's important to note that this method may not cover all internal customers, especially if the catalog is outdated.

Conducting internal customer surveys is another practical approach. You can gain insights into their needs and expectations by gathering direct customer feedback. However, there is a potential downside: surveys may have a low response rate, leading to incomplete data and potentially skewed results.

Holding **cross-functional meetings** with representatives

from different departments is a great way to build relationships and improve communication. These meetings can help you uncover IT service needs and identify internal customers. However, they can be time-consuming and may not cover all internal customers.

IMPLEMENTING a help desk ticketing system is another way to track IT requests and identify the departments or teams making those requests. This approach provides detailed data on IT service usage. However, it's worth noting that it may not capture all internal customers if some individuals or departments do not use the help desk.

If you're willing to adopt new tools and processes, implementing collaboration tools can be beneficial. These tools can track interactions between IT and other departments, helping you identify internal customers. The advantage here is that you can access real-time data on interactions. However, be prepared for the potential challenges of adopting new tools and processes.

Reviewing how your IT budget is allocated across departments can provide valuable insight into resource allocation and help identify internal customers. However, remember that this method may not accurately reflect all IT service usage or needs.

Monitoring the usage of IT assets, such as hardware and software, can also help you identify which departments or teams are using these resources. This method provides accurate data on resource usage. However, it's important to note that it may not cover all internal customers, especially for shared resources.

NOW, let's discuss ideas for identifying your external customers.

One approach is to **review customer complaints** first, followed by compliments. This allows you to collect direct customer input and easily quantify and analyze the data. Additionally, this method has the potential to reach a large audience. However, customers may not always be truthful or provide detailed responses, and response rates may be low.

Organizing focus groups can provide in-depth, qualitative insights into customer needs and expectations. These groups encourage open discussions and brainstorming sessions, allowing customers to provide real-time feedback. However, remember that focus groups have a limited sample size and can be time-consuming and expensive to organize.

One-on-one interviews provide a personal and in-depth understanding of individual customer needs. These interviews also help build customer rapport and trust and encourage honest feedback. However, conducting one-on-one interviews can be time-consuming, and difficult to generalize findings.

Analyzing customer support data is another valuable method. By utilizing existing data, you can identify trends and pain points, which can then inform continuous improvement efforts. Additionally, this approach allows you to prioritize areas for improvement based on real-time feedback. However, it's important to note that relying solely on customer support data may not cover all customer needs or expectations, as it may be skewed based on the type of customer contacting support.

Collaborating with industry influencers and thought leaders can provide valuable insights into customer needs and expectations. Leveraging their expertise and audience can enhance your credibility and brand recognition. However, collaborations with influencers may be expensive, and their opinions may not always align perfectly with your company's goals.

Lastly, **creating customer advisory boards** can establish long-term relationships with key customers and provide access to ongoing feedback and insights. This approach encourages customer loyalty and advocacy. However, it's important to note that creating and managing a customer advisory board requires dedicated resources and may not represent the broader customer base.

These are just some of the many strategies you can use to identify your internal and external customers. As you embark on this journey, carefully consider each method's pros and cons and tailor your approach to your organization's specific needs and goals.

Remember, due diligence is vital when it comes to understanding your customers and meeting their needs.

Using these pointers will likely gain significant support and promotion from your internal customers, such as employees, managers, and other organizational stakeholders. All of this is essential for ensuring the growth and usefulness of a CRM, but don't assume these are the only things that need to be done. You must "read to room" and continually return to these ideas. Note: the second you stop caring, your internal and external customers will too. By actively engaging internal customers and fostering a culture of collaboration and innovation, organizations can maximize the benefits of their CRM and drive its ongoing success.

Ideas for Internal Customers (and Users)

MOST IMPORTANTLY, you can't fake this stuff. If leadership or you don't care, you won't be able to fake it, and people (internal users) will notice, and their support or energy will disappear. This should scare you if either of these two groups don't care - this situation must change; if you need motivation, then here are some strategies and idea to think about - One of the key strategies is to encourage internal customers to promote and support the CRM is to **communicate the benefits and value** it provides effectively. This involves clearly articulating the advantages of the CRM, focusing on the features and capabilities most relevant to their needs and goals. Demonstrating how the CRM can improve efficiency, streamline processes, enhance customer relationships, and ultimately contribute to the organization's overall success helps internal customers understand its importance and encourages them to support and promote it actively.

Providing **comprehensive training and support** is another crucial aspect of engaging internal customers. By offering personalized training sessions and ongoing support, internal customers can build confidence in using the CRM and develop a deep understanding of its features and capabilities. Equipping them with the knowledge and skills they need to use CRM effectively empowers internal customers to become advocates for the system and encourages them to share their experiences and insights with others.

Involving internal customers in decision-making is also essential for gaining their support. Engaging them in the CRM's planning, implementation, and ongoing management provides opportunities for them to contribute their ideas, feedback, and expertise. This involvement helps internal customers feel a sense of ownership and responsibility for the CRM's success, which can motivate them to support and promote it actively. Active collaboration and innovation is another critical strategy for encouraging internal customers to promote and support the CRM. It's essential to have an environment where employees can collaborate, share ideas, and collaborate on CRM-related projects and initiatives so that internal customers can feel more connected to the CRM and its objectives. This connection inspires them to support and promote its growth and usefulness actively.

Recognizing and rewarding success is essential for creating a positive atmosphere within the organization and motivating internal customers to continue championing the CRM. **Celebrating the achievements and progress** of those who actively support and promote CRM reinforces the value of their efforts and encourages continued engagement. By recognizing and rewarding success, organizations can help create a sense of pride and commitment among internal customers.

Establishing clear expectations and accountability is crucial for encouraging internal customers to actively promote and support the CRM. By setting specific goals, deadlines, and performance metrics and holding internal customers accountable for meeting these expectations, organizations can provide the necessary structure and support for their employees. Providing regular feedback and assistance helps internal customers stay on track and address any challenges they may encounter.

Gaining the support and promotion of internal customers is essential for the growth and usefulness of a CRM. By actively engaging internal customers, communicating the benefits and value of the CRM, providing comprehensive training and support, involving them in the decision-making process, fostering a culture of collaboration and innovation, recognizing and rewarding success, and establishing clear expectations and accountability, organizations can better ensure the ongoing success of their CRM and maximize its benefits.

The non-technical underpinnings of embracing a CRM lie in the belief that it can fundamentally transform how organizations interact with their customers, streamline their operations, and ultimately contribute to their long-term success.

One of the core reasons for adopting a CRM is its potential for **enhancing customer-centricity** within the organization. At the heart of every successful business lies a deep understanding of its customers' needs, preferences, and expectations. By using a CRM, leaders and internal customers can gain valuable insights into customer behavior, preferences, and trends, guiding their decision-making and enabling them to serve their customers better. This can help to increase customer satisfaction, loyalty, and advocacy, all of which are critical for the sustainability and growth of the organization.

Another reason for embracing a CRM is the belief in the **power of data-driven decision-making**. In today's fast-paced and competitive business environment, organizations that can effectively leverage data to inform their strategies and operations are more likely to succeed. A CRM provides a wealth of data on customer interactions, sales, marketing efforts, and more, which can be harnessed to make informed decisions and optimize business processes. By adopting a CRM, leaders and internal customers demonstrate their commitment to evidence-based decision-making and continuous improvement, values that are essential for fostering a culture of innovation and adaptability.

Moreover, adopting a CRM is a testament to the organization's commitment to collaboration and breaking down silos. A CRM enables seamless information sharing and cooperation among different departments by creating and properly operating a centralized platform for storing and accessing customer data. This can lead to a more holistic understanding of customers, improved cross-functional communication, and improved decision-making and problem-solving. Embracing a CRM, therefore, signals a willingness to work together as a cohesive unit, putting the needs of the customer and the organization above individual or departmental interests.

The argument for adopting a CRM also extends to personal growth and development. By using a CRM, leaders and internal customers can develop new skills, expand their knowledge, and better understand their customers and industry. This continuous learning and growth mindset not only benefits the individual but also contributes to the overall success and competitiveness of the organization. By adopting a CRM, leaders and internal customers demonstrate their commitment to personal and professional development and their belief in the potential of technology to enhance their capabilities and effectiveness. Adopting a CRM can be seen as an expression of the organization's values and vision. By investing in a CRM, organizations signal their commitment to customer satisfaction, innovation, collaboration, and continuous improvement. These values, in turn, can inspire and motivate leaders and internal customers to use the system as they recognize its alignment with their own beliefs and aspirations. By appealing to these shared values and vision, you should realize unity among employees, further driving CRM adoption and effective use.

The actual reasons for adopting a CRM extend far beyond its practical and technical benefits. By appealing to the intrinsic motivations of leaders and internal customers, organizations can inspire them to embrace the CRM to enhance customer-centricity, foster data-driven decision-making, promote collaboration, support personal growth and development, and embody the organization's values and vision. By doing so, organizations can ensure their CRM's ongoing success and create a positive and supportive culture that drives innovation, adaptability, and long-term growth.

Comparing Internal And External Customers

SINCE INDIVIDUALS BEGAN CREATING and selling goods, external clients have been a part of the business. Internal customers, on the other hand, are a more recent concept.

Kind and compassionate leadership, fair and equitable remuneration, pleasant conditions for workers, and cutting-edge technology are all examples of a healthy work environment. The theory holds that the stronger employee morale is, the more loyal and productive those people are.

That sense of belonging can also come from the knowledge that they contribute to something bigger, which benefits their operation. Satisfied internal consumers are linked to satisfied external customers.

However, aside from the feel-good factor, using "internal customers" may have certain practical advantages. IT can assist teams in prioritizing problems and schedules to increase inter-departmental communication for groups that operate with internal and external clients.

It's easy to notice how the former's experience can affect the latter's after you've decided to handle customers as internal and external.

For example, if your internal customers work in a call center and interact with external customers, the tools they rely on must be reliable and easy to use. Suppose your employees are dissatisfied with outmoded processes and technology. In that case, that dissatisfaction will most likely be mirrored when your external clients want assistance, resulting in a double whammy—not a good formula for success.

Internal Customers Of An Organization

IN RECENT YEARS, companies have realized the importance of providing a good customer experience (CX). Around the world, businesses have taken the initiative to thrill their customers and convert followers into ardent supporters. But what about your staff? Who are your internal customers? Internal customers can either be your best friends or your worst nightmare. They work for your organization and need help from the CRM (or at least the outputs from a CRM). They can be in various roles, from "fingers on keyboards" to division or geography leaders. As a person focused on CRM success, you must treat all internal customers very seriously and never pass over a team just because they have only a few users and/or because another team of internal users is making more noise. Grade internal users by the decisions they must make, the quality and importance of the information they can provide, and where they sit in the workflow(s) you are trying to solve.

FOR EXAMPLE, the person who sends emails to a social media team relies on editors and illustrators to compose and design the emails. Because the coder was rushed, the email may get delayed or published with mistakes if the writer runs late with the content to the coder. Internal consumer relationships include a lot of peer-to-peer connections.

Not Taking Care Of Your Customers

LET'S take a moment to reflect on the realities of CRM implementations. Despite their numerous benefits, the failure rate of CRMs has been alarmingly high since their introduction years ago. In my experience, much of this can be attributed to what I call "not taking care of your customers." You see the success of a CRM hinges on **meeting the complex and multifaceted needs of internal and external customers**. While we may have expected the failure rate to decrease as the CRM market matured and products improved, the truth is that the reasons for CRM project failures continue to persist.

After a few years post-implementation, many CRMs fall short of expectations, leaving organizations with subpar performance and little return on their investment. It is disheartening that leadership often fails to acknowledge the impact or returns brought by the CRM.

Now, I'm well aware that many of the problems and fixes in CRM are not exclusive to this realm; they can be found in various IT software projects. However, CRM is unique because it is relatively new and requires active group participation and sharing. Even with the adoption of agile growth and best practices like ITIL, software implementations in the CRM space remain risky undertakings, often leading to failure.

But here's the critical insight: No business can achieve its objectives without maintaining the correct relationship with its leads or customers. That is the heart of CRM. Over the last few decades, we've witnessed a significant shift in customer relationships, bringing new challenges and opportunities. And now, we can explore those changes and discuss practical strategies for nurturing and maintaining a healthy customer relationship.

Success in CRM comes not from treating it as a mere software implementation but from understanding the fundamental importance of building and nurturing customer relationships. Let's make this journey together, armed with wisdom and a commitment to putting our customers first.

The Nature of Customer Relationships is Changing

THE AGE **of Social Media**

In this social media-driven world, customer relationships have changed dramatically in the last decade. Once businesspeople discovered that there is a massive possibility of engaging with customers on social media that can be utilized to make stronger bonds with each customer, the dynamics of brands and their relationship with customers changed dramatically.

If customers have a negative experience with a product or service, they can quickly vent their frustrations on social media, tarnishing the company's positive reputation and image. Businesses take customer relationships more seriously because of the risk of spoiling the brand's image.

This has resulted in a massive rise in social media interactions between customers and businesses. After all, negative consumer feedback could result in a significant loss for the company, and no company can afford to sit back and let that transpire.

Customers are always more enthused to speak their minds on social media because of its frictionless nature. As a result, when customers receive negative feedback from a company's support team, they immediately seek out the simplest means to raise knowledge or share their experience.

As a result, companies are now engaging with customers more frequently by interacting with them on social media to reduce the damage to the customer relationship and the brand image. 24/7 chatbots and chat boxes with human customer service agents sitting on the other end. This is being done. Other brands have active social media pages, where they actively respond to customer queries and concerns no matter how fucking deranged the issue. Identifying the problem and assuring them that it will be resolved as soon as possible helps to restore a company's reputation and build a strong relationship with customers.

8

The Wrong People Are Everywhere - Identifying Them

I have been waiting to write this section from day one. This topic is the single catalyst for my writing the entire book.

THE "WRONG" people can also customers, usually internal customers, but they can be external too. The "wrong" people are everywhere, and they are critical to the outcome of the CRM. I believe this to be true for the following reasons:

WARNING: This is the single most essential AND Overlooked reason that CRMs fail (IMO). Identifying these people/divisions/geographies as quickly as possible is amazingly important. The disruptions caused by saboteurs and assholes cannot be allowed to infringe further and slow down CRM advancement. After speaking with and being one myself, so many professionals have been bullied and beaten up by these people it is now fun to dig in the mud and sling it back.

INTENDING to frame our discourse around the pivotal role of CRM in businesses, we need to commence this chapter from a place of shared understanding. We are operating under certain assumptions that form the basis of our multifaceted discussion. The hypotheses that guide our narrative revolve around familiar settings of conducting business and an array of personalities that participate in it.

The preliminary assumption that is fundamental to our deliberations is that our context remains grounded within the sphere of typical business environments. I assume you're not operating in an extraterrestrial setting or dramatically unconventional circumstances. So I also know you have saboteurs and assholes around you. 😱 These characters are woven into the fabric of this narrative and are those you work with, commonly referred to as staff, peers, and leaders.

Every member of a business ecosystem plays a vital role in shaping the dynamics of the organization. Staff members are the backbone for executing tasks and keeping things running smoothly. Peers bring diverse skills and perspectives, offering intellectual stimulation and motivation. And leaders drive the business forward, making important decisions and guiding the organization toward its vision. But within each role, some unique opportunities and challenges shape the operational landscape. The varied agendas, perspectives, and motivations of individuals in these roles create a nurturing and competitive environment. It is this vibrant mix that forms the foundation of my belief.

In simpler terms, every person in a business has a job to do. Staff members get things done, peers bring new ideas and motivation, and leaders make the big decisions. But each role also comes with challenges and opportunities, and the diverse mix of people in the organization can create a dynamic and sometimes competitive environment. It is this energy that is crucial to the success of any business.

The second assumption deals more with human interactions within the business premises. Despite an idealistic view propagating harmony and shared goals, the reality of any workplace encompasses a spectrum of attitudes (sometimes the worst being stubborn jerks) and behaviors (sometimes being selfish and self-serving assholes). In this spectrum, it's pretty plausible that you'll encounter individuals with whom you don't share a comfortable relationship.

Not everyone in a company will be friendly or exhibit professional courtesy. - crusty assholes. There are various reasons why this might be the case. It could be due to personal biases, a competitive nature, pressure, stress, or individual personality traits. A typical workplace's many professional and emotional tensions further complicate these diverse human interactions. Given the range of interpersonal dynamics you'll encounter, it's essential to establish solid systems that ensure smooth operations regardless of these human quirks. It's not an easy task, but it can be done. This is where CRMs come in. They provide a robust platform that can transcend personal dynamics and help businesses maintain a customer focus, even when dealing with complex individuals in the workplace.

Navigating this complex human puzzle requires careful management of expectations and personal boundaries and preserving established hierarchies. Successfully implementing and effectively using CRMs is not just about understanding the technology. It's about navigating the unseen human dimensions and power structures and getting the system recognized and accepted among the people involved. It takes effort and a strategic approach but is essential for long-term success. These assumptions may seem diverse but are standard and expected in business. Experienced professionals would agree that these dynamics are subtly intertwined in every organization's fabric. When technology and human interactions converge, these inherent patterns become more pronounced. Recognizing and accepting them is the initial stride towards navigating and effectively dealing with them, whether by collaborating harmoniously, finding workarounds, or confronting them head-on.

Understanding the importance of integration for CRM Solutions is vital to all involved. Things are changing rapidly, some people are slowing down from their original cruising speed, and others have completely stopped in the business world, and within the technology it supports. It's important to grasp a few essential concepts regarding CRM solutions because they're not meant to exist in their isolated technical bubble. They should be integral to the more comprehensive business structure, fitting seamlessly within people and operations. These ideas paint a realistic and crucial picture of your professional environment, which is particularly important when navigating the world of CRM adoption. The link between business success and effectively utilizing CRMs is well-documented. However, achieving success in this context goes beyond technological triumph alone. It also relies on how well the solution becomes ingrained within the organizational ecosystem. This integration requires a keen understanding and careful navigation of the operational landscape and its unique characteristics, including the intricate web of human dynamics that permeate the professional sphere.

Setting The Baseline

TO HAVE AN IN-DEPTH CONVERSATION, let's establish some baseline assumptions. While they may seem obvious, they are crucial to understanding CRM dynamics.

Let's assume everyone is embedded within the same daily orbits of business life at a company/firm/organization and can understand different perspectives as they are presented. You navigate interactions with peers, staff, and leaders within your organization. Each individual plays a distinct role and has their influence, collectively forming the organization's lifeblood.

The CRM is vital in driving customer-centric strategies in modern businesses. **However, it is met with varied responses across the organization.** Some individuals may resist its introduction or feel unsupportive of its ongoing operation. Understanding these diverse reasons will give us valuable insights into the human element of CRM adoption.

Unraveling Resistance

AS MENTIONED before - Resistance to CRMs often stems from a sense of threat. This can arise from a **fear of change, concerns about transparency, or anxiety over skills inadequacy**. Employees may worry that the new system will disrupt their routines and upset their comfort zone. On the other hand, some individuals prefer to keep their work activities private and may feel uncomfortable with a CRM's ability to document every action. Additionally, employees may be apprehensive about mastering a new tool, feeling vulnerable in the face of its introduction.

Another catalyst for resistance is our natural aversion to change. Introducing a new system, like a CRM, disrupts the established order, prompting individual discomfort. They may express frustration at the added effort required to adapt to the new system, pushing back against its implementation.

WARNING: **This is scary, but important stuff - Political undercurrents** in organizations can further hinder CRM acceptance. Politics often influence perceptions and actions, steering decision-making from rationality to subjective power-oriented motives. If someone in a management position views the CRM as disadvantageous to their aspirations, they may use their influence to undermine the initiative.

PARALLEL to this political dimension is the practice of alternative hierarchy or empire-building. Some individuals try to establish power centers within the organization. A highly transparent CRM that standardizes processes and distributes authority evenly may be seen as a threat to their "kingdoms," leading to resistance.

In some cases, competing ideas or concurrent initiatives within an organization can create barriers to CRM acceptance. Individuals may advocate for alternative solutions if they believe another process or tool can achieve similar results. Additionally, introducing an advanced CRM might overwhelm the transition if the organization undergoes multiple changes.

Empathetic Understanding and Tailored Solutions

IT'S essential to shed light on these factors without assigning blame. We aim to understand the varying perspectives co-existing within the organization empathetically. Leaders can effectively tailor the CRM implementation process to address these fears by recognizing these feelings.

For example, creating a culture of learning and support can alleviate concerns over the skills deficit caused by the new system. Fostering a change-embracing and transparent environment can mitigate anxieties about transformation and openness. Moreover, ensuring a fair and inclusive decision-making process will prevent politically motivated interference. Appreciating and collaborating on alternate solutions can also neutralize resistance from competing notions.

Understanding and addressing these dynamics is crucial for successful CRM implementation and operation. Each organization has a unique character and inherent dynamics, presenting varying challenges requiring customized solutions. We will explore strategies that facilitate a seamless CRM transition, enhancing its acceptance and efficacy in organizational scenarios.

9

Identifying Sticks In The Mud

We will help those who don't think these people don't exist or don't recognize the problems they will create. They do, and they will cause more harm than your champions will create opportunity and goodwill. Sticks-In-The-Mud (SITM) are different from saboteurs and assholes. They (SITM) are just reluctant to change. They primarily don't have an ill will on their mind or try to flip your game table. More than likely, they are resistant to change and/or skeptical of the promised results. Despite this difference, the outcome of their lack of utilization of the CRM can be just as damaging as saboteurs and assholes. So they must be converted if your CRM effort is to succeed.

As we continue to discuss the details of CRM implementation, it is impossible to do so without addressing the inevitable human element that permeates the process. Understanding and navigating human perspectives, particularly those that conflict with the organizational objectives surrounding the CRM, are as critical to the success of the CRM project as the technology and process improvements themselves.

One of the key challenges organizations face is identifying and addressing individuals who may be skeptical or wary of the CRM initiative. These individuals might initially pose resistance to the CRM effort or, worse still, morph into what we colloquially call 'nemeses' – individuals who secretly work against the CRM effort with a façade of your teams or the more significant public, corporate support. Dealing with these nemeses is essential to ensure the seamless integration and beneficial operation of the CRM within the organization.

To effectively engage these wary parties, converting them into champions of the CRM should be a central goal. The support of newly converted champions is indispensable, sometimes carrying as much weight, or even more, as the endorsement of existing proponents. Such transformation reaps twofold benefits, reducing resistance on the one hand and shoring up support on the other. Through this metamorphosis, a previously skeptical individual can become one of the initiative's most ardent advocates, with a greater appreciation for the system owing to their transition journey.

However, it is crucial to recognize that approaching this challenge may not be straightforward. Recognizing and acknowledging resistive forces is not always a given; occasionally, organizations remain oblivious to such factions. Failing to identify nemeses or recognize the potential damage caused by their actions not only hinders the CRM initiative but also undercuts the positive contributions of the project's champions.

The apt management of these nemeses, be they overtly antagonistic or covertly undermining, requires a precise and deliberate strategy. Once identified, the critical decision lies in the approach taken to address their concerns or actions. Blatantly confronting or reporting them to superiors may not always yield favorable outcomes and might instead exacerbate matters. Conversely, engaging them in dialogue to understand the rationale behind their resistance could yield valuable insights and open up avenues for constructive collaboration.

There are two principal ways to handle nemeses effectively.

- Firstly, hope for their eventual separation from the organization, which, while tempting, may not be the most practical or likely occurrence. (We all wish this secretly a lot, I think) 😎
- Secondly, and perhaps more realistically, Figure out a customized and personal way to train and convince them to change their behavior and stance. They likely strive to comprehend any project or process they hold dear and explore the fears around how the CRM might jeopardize it.

Ask yourself, why are they stuck in the mud? What is driving the resistance?

By empathizing with their concerns and identifying the specific aspect that fuels their apprehension, leaders can build solutions that incorporate and bolster these elements using the CRM. Leaders should involve these individuals, ensuring they feel a sense of ownership and pride in mutually enhanced outcomes.

Organizations activate these individuals' intrinsic motivation by offering them greater responsibility for the system's success in areas they are passionate about, paving the way for positive affirmation. Publicly appreciating their efforts is an essential reinforcement that bolsters their commitment.

The human element within the CRM implementation process is undeniably complex, comprising differing perspectives and motivations. The objective is not to vilify the nemeses but to understand and address their concerns while channeling their energy toward a positive outcome. Engaging them in meaningful conversations, empathizing with their fears, incorporating their valued elements into the CRM solution, and assigning them significant responsibilities can help win them over.

When exploring the world of CRM adoption, it's essential to

examine the intricate layers of human factors that come into play. This chapter delves deep into these factors, unraveling the threads of securing buy-in from leadership, addressing staff concerns, and fostering collaborative relationships within the CRM ecosystem. Above all, it sheds light on how CRM solutions can be positioned as a sacred and secure conduit for valuable information, fostering trust and efficiency among all parties involved. Amidst the complex maze of human emotions and operational intricacies, CRM solutions emerge as more than technological tools. They become an essential entity intricately linked to the people and processes within your organization.

With much insight and a little luck, the upcoming sections aim to help you deal with complex individuals who may pose challenges to your projects or programs. Often, people with abrasive personalities are the ones to consider first when evaluating the dynamics of your team. In most cases, they can't help themselves. It's unfortunate, but it's worth noting that a significant portion of this group tends to be highly experienced professionals nearing the end of their careers. These individuals sometimes have "no skin in the game," and their reference points often harken back to when modern systems didn't have the capabilities they hold today. For them, sharing information may be seen as a threat to their perceived wisdom and secret knowledge. On the other hand, the systems we are discussing thrive on a community spirit, valuing open information sharing and collaboration.

The transformation of doubters into enthusiastic champions of a CRM effort can significantly enhance the value of the overall initiative. When these initially skeptical individuals lend their support and actively advocate for the project, it adds credibility and positivity that can counter any lingering resistance. It's truly remarkable to witness their conversion journey. These former skeptics become some of the most powerful CRM adaptability and value proposition demonstrators.

Their transformation sends a clear message: the CRM can win over even the most skeptical minds. When those who initially doubted the system's potential become its most ardent supporters, it serves as a powerful testament to the transformative capabilities of the CRM solution. This kind of endorsement can inspire others to believe in the system, fostering a sense of confidence and enthusiasm throughout the organization.

Remember, the journey of transforming skeptics into champions is not always easy. It requires patience, effective communication, and a genuine understanding of their concerns and reservations. However, the rewards are well worth the effort. By embracing these individuals as advocates, you strengthen the CRM initiative and create a positive ripple effect that can propel the organization toward tremendous success. So, don't underestimate the power of turning skeptics into believers—it can genuinely transform the potential of your CRM and elevate the entire organization.

10

Understanding "Challenging People"

The Importance of Identifying Difficult Personalities

IT'S no secret that customers are the heart of success. However, the journey of implementing a transformative tool CRM can be a challenging one. An often overlooked aspect can significantly impact the outcome: **the human factor**.

The identified people (users) using the CRM can make or break the implementation process. While many organizations understandably focus on the technical aspects of CRM implementation, such as choosing the right vendor and setting up the software, an equally important consideration deserves our attention. We must proactively identify and address any potential problems arising from introducing a new system into an environment where employees are already dealing with many stressors, including difficult personalities.

Undoubtedly, when it comes to implementing a CRM, the technical challenges are just the tip of the iceberg. While factors such as system integration, data migration, and alignment with existing business processes are essential considerations, we must also prepare ourselves for the potential hurdles that some may call "difficult" personalities within our organization. These individuals, whom I kindly refer to as "challenging" individuals, can present an array of obstacles beyond the technical aspects of CRM implementation.

From my years of hands-on experience, I've witnessed the detrimental impact that these personalities can have on CRM projects. They are like a fly in the ointment, spreading negativity, resisting change, and stubbornly clinging to outdated methods. Their negative attitudes can shadow the entire implementation process, making fostering a culture of acceptance and enthusiasm difficult.

To successfully navigate the rough waters of dealing with challenging individuals, it is imperative to identify and understand them before launching a CRM. Take the time to shine a light on their concerns, fears, and reservations. These individuals often harbor legitimate worries, such as a fear of losing control, job security concerns, or doubts about the new system's effectiveness.

You stand a better chance of winning their support by proactively addressing their apprehensions and engaging them in the process. Empathy, communication, and education are your secret weapons in dealing with these challenges. Take the time to listen, empathize, and explain the reasons behind the CRM implementation. Show them how the new system can make their work more accessible and efficient and ultimately contribute to the organization's overall success.

In my experience, resistance often lessens when people feel heard and understood. By involving challenging individuals in decision-making and giving them a sense of ownership, you can transform them from adversaries to advocates. Leverage their unique perspectives, and turn their negative attitudes into a driving force for positive change.

Remember, that dealing with challenging individuals is not a one-size-fits-all endeavor. Everyone is unique, and your strategies must be tailored to their concerns and communication styles. Building a coalition of support requires patience, persistence, and adaptability.

So, as you embark on the journey of CRM implementation, don't underestimate the impact of complex individuals. Anticipate their existence, employ strategies to address their concerns, and forge a path that will win over their support. Doing so will ensure a smoother implementation and set the stage for broader organizational success. Remember, it is through challenges that we grow and learn.

So, why is specifically acknowledging such characters before a CRM launch crucial? - By recognizing these individuals and their potential impact on the implementation, we can plan carefully and adopt a tailored approach to ensure their involvement and buy-in. This early recognition sets the stage for a more fruitful CRM adoption.

During the CRM launch, we'll explore the significance of acknowledging and understanding these complex personalities. We'll delve into specific strategies to overcome resistance and nurture a positive environment that promotes the successful adoption of your CRM implementation.

Remember, a comprehensive understanding of the human factor is critical to achieving CRM success. By acknowledging the challenges posed by difficult personalities and proactively addressing them, we can carefully navigate the implementation process and ensure everyone is on board for a smoother journey toward CRM excellence. There's much more to learn, and I'm here to guide you through the intricate path of CRM implementation. We'll overcome challenges and create a customer-centric culture that nurtures lasting success.

In the realm of CRM, a little foresight and consideration for the human factor go a long way.

11

Not Knowing What The Company Wants

-THE TRUE GOALS-

Determine What Your CRM Needs To Do

Don't get all crazy here; pondering a super-intergrated mega-system for all but the biggest companies in the world will not work. You will have a thinly veiled, watered-down pile of crap with Swiss cheese holes. Having a detailed and laser-focused system for the most essential parts of your business is still extremely important. The best way to ensure this happens is by picking 2 or 3 areas and developing them to their fullest potential. So what are the most critical areas of your business?

- **Sales:** A detailed sales process means you can train new team members and ensure everyone is doing things as they should. It also allows you to scale up quickly when needed (which will happen).
- **Customer Service**: This should be an afterthought and something built into every part of your business. Whether dealing with returns or complaints, having a clear set of steps makes it easy for anyone to follow them to solve any problem. -Operations: This is ensuring that

everything that needs to happen happens. It includes things like payroll and finances, systems for shipping and receiving products, and customer service issues. -Sales & Marketing: Having a clear set of steps for how your sales team should be working will ensure that they are always on track and focused on what matters most (which is why it's essential).

- **Finance:** This is one of the most critical areas of any business and needs to be handled by someone who knows what they're doing. Having a detailed system in place will assist you in keeping track of all your finances so that you know where your money is going at all times.
- **Product Development:** How do you ensure your products are the best they can be? This is an essential question, and having a product development process will help you ensure that everything is done correctly.
- **Marketing:** Your marketing strategy must be clear, concise, and easy to follow. Your team members need to know precisely what they should be doing at all times for all of your efforts to work together seamlessly. This is the most crucial part of your business but is also the hardest to get right. By having a detailed marketing process, you can ensure that everyone is doing things the right way. This will save you time and money down the road.
- **Finance:** This is the one that most people forget about. But you risk losing it instantly if you don't have a money management process. By setting up clear rules and guidelines for keeping track of things like accounts receivable and payable, you can ensure that your business has a solid foundation to build.
- **Flexibility Of Functionality**: The number of apps available and the ability to build customized features must be considered. After the setup process, you would want to make improvements, and if this is complicated, it will become a significant expense down the road.

- **Integrations:** You'll almost certainly need to bind your CRM to other systems you use, and it's important to know your CRM's core functionality immediately. The more versatile, the better, and the simplified CRM to integrate would significantly reduce customization costs.

All this creates tremendous benefits and costs, but flexibility is crucial. The more feature-rich your system, the more it will cost to implement and maintain. Considering these factors and others, use them to make the most informed choices about using a cloud-based CRM or staying on your own server.

If you are trying to build a system that does everything, you will be stuck in an infinite loop of one department wanting something requiring a trade-off in other departments. If you can build/buy a system that completely solves one workflow and then provide gateways so that other departments can plug into it, then you will have done much better than if you had tried to put every single department's entire workflow into your system out of the gate. But equally alarming is if you select a specific individual workflow and the solution only solves 80% of the problems people have on 100% of their issues. In short, they will ultimately be upset if they still have a complex process solved due to the new CRM "fix" and thus still needs manual babysitting. The failure will be broadcast through the rumor mill and spread like wildfire.

Any new software tool can be enticing, but there's something particularly alluring about technology aiming to solve a long-standing issue. Many CRM products claim to optimize the intricate and often elusive process of weeding out the *barnacles* (customers with low loyalty and small margins) and attracting *true friends* (customers with high commitment and big margins).

And while CRM can do that, it must be backed by well-thought customer segmentation and targeting strategy. You can't ask the software to look for something but not specify what you want.

But why should you formulate a customer strategy and spend on CRM software?

Well, the answer is simple, of course. It would be best if you had a concrete customer strategy to have a marketing goal to work towards. Good old-fashioned segmentation research is the foundation of successful CRM. Furthermore, it is tailored to meet unique marketing objectives.

Implementing CRM without first performing analysis/interviews with VIPs and assessing marketing objectives - Note: This effort will fail if you assume you know the problems before you ask everyone what their issues are and listen with focused attention. Despite this, many executives confuse CRM technology with a marketing campaign. So, they give complete control of planning their marketing objectives to the vendor sourcing their CRM software.

Alternatively, companies buy a CRM from a vendor and then reverse engineer the process to formulate a customer strategy based on the technical capabilities of the particular CRMs specifications.

Identifying and deploying a successful project

ARE you looking for a new feature for your CRM? Make sure you know what it can do natively first. You may have heard about how great specific add-on tools are and how valuable they will help you, but do not assume that every company that sells a CRM has your best interests in mind. Many of these companies)including the mega-CRM companies themselves) are more interested in making money from their products than helping you succeed with them. - Training and Development, Inconsistent Leadership.

It's common, unfortunate, and still fascinating to watch companies plan and launch CRM projects; months later, I often look at their implementations and figure out why they failed.

Identifying a small and low-hanging fruit project within your CRM can sometimes be a great way of showing not convinced leaders and customers that the system can work for their specific needs, streamlining processes, and creating benefits for the entire company. The tricky part of this strategy is not getting caught up in making too many of these efforts, as they frequently spiral off into many directions and can set the expectations of other leaders and customers that they will also get the same service. Beware that going down this path is easily a path of certain destruction and will cause unhappy customers, as you will not be able to support many one-off solutions. You will experience a high turnover of your CRM support staff.

The philosophical challenge of placating difficult leaders and customers in the context of a CRM lies in striking a balance between addressing their specific needs and concerns while maintaining a unified and consistent approach that serves the broader interests of the organization. This requires a deep understanding of these stakeholders' underlying motivations and expectations, as well as the ability to communicate and demonstrate the value of the CRM in a way that resonates with their unique perspectives and priorities.

One of the critical philosophical principles in navigating this challenge is the concept of empathy. One can better understand their concerns, frustrations, and aspirations by putting oneself in difficult leaders' and customers' shoes. This empathetic approach can help identify the root causes of their resistance to the CRM and reveal opportunities for addressing their needs without compromising the overall integrity and effectiveness of the system. Empathy also fosters a sense of trust and rapport between the CRM support team and these stakeholders, which can pave the way for more constructive and collaborative relationships.

Another essential philosophical consideration is the notion of flexibility and adaptability. While it is crucial to maintain a consistent and coherent approach to implementing and using a CRM, it is equally important to recognize that no one-size-fits-all solution exists. Each organization, department, and individual will have unique needs and preferences that may require some degree of customization or accommodation. By adopting a flexible and adaptable mindset, CRM support teams can work with difficult leaders and customers to identify mutually beneficial solutions that address their specific concerns while preserving the overall goals and objectives of the CRM.

The principle of patience and persistence also plays a crucial role in dealing with difficult leaders and customers. Change is often met with resistance, and the adoption of a CRM is no exception. It may take time for these stakeholders to fully appreciate the benefits and value of the system, and CRM support teams need to remain patient and persistent in their efforts to demonstrate its merits. This may involve providing ongoing support, guidance, and reassurance and celebrating small victories and progress. By remaining steadfast in their commitment to the success of the CRM, CRM support teams can gradually win over even the most skeptical and resistant stakeholders.

In addition to empathy, flexibility, and patience, the philosophical approach to placating difficult leaders and customers must also encompass the principles of transparency and accountability. By openly and honestly communicating the rationale behind the CRM, its intended benefits, and the potential challenges and limitations, CRM support teams can foster a sense of honesty with these stakeholders. This transparency can help dispel misconceptions and alleviate concerns while setting realistic expectations for what the CRM can and cannot achieve. Moreover, by holding themselves accountable for the success of the CRM and demonstrating a willingness to learn from mistakes and make improvements, CRM support teams can further engender confidence and goodwill among difficult leaders and customers.

Another thing to consider is that an approach to dealing with hard-to-make-aware leaders and customers should be grounded in collaboration and shared ownership. Rather than viewing these stakeholders as adversaries or obstacles to be overcome, CRM support teams should seek to involve them in the decision-making process and empower them to contribute their ideas, insights, and expertise. This collaborative approach can create a sense of shared responsibility for the success of the CRM and foster a culture of innovation and continuous improvement. CRM support teams, difficult leaders, and customers can forge stronger relationships and achieve better organizational outcomes by working together towards a common goal.

Making difficult leaders and customers happy in the context of a CRM requires a non-technical and technical approach rooted in empathy, flexibility, patience, transparency, accountability, and collaboration. By adopting this mindset and applying these principles, CRM support teams can navigate the delicate balance between addressing these stakeholders' specific needs and concerns while maintaining a unified and consistent approach that serves the broader interests of the organization. Although this path may be fraught with challenges and potential pitfalls, it is ultimately a journey that can yield significant rewards for enhanced customer relationships, streamlined operations, and long-term growth and success.

12

Case Study: Difficult People Spreading The Wong Ideas

Scene:

Following an enthusiastic meeting where all division leaders agreed on implementing new reports and dynamic dashboards, the CRM team eagerly deployed the new system to all groups across the globe. It appeared that everyone would benefit from this improved CRM infrastructure, and it was expected to enhance operational efficiency exponentially.

Incident:

Weeks later the CRM team was surprised to receive numerous support tickets from a specific division shortly after the deployment. Upon investigation, it was discovered that the dissatisfaction arose from staff at several levels were being coerced to use the new CRM system, which they saw as a redundant and meaningless task. The

Reason? Further investigation uncovered that the leader of the disgruntled division had indeed forced everyone to use the new CRM system despite still wanting to see "his" old reports. This resulted in double the division's workload, negatively impacting morale.

Result:

Despite the new reports and dashboards being fully operational, they became virtually worthless due to poor leadership decisions and a reluctance to adapt to newer, and agreed upon, technologies. What is even worse, the leader's peers in other divisions soon followed suit, leading to widespread frustration and further casting doubt on the new CRM system. This one poor decision significantly impacted data quality. It caused tremendous damage, requiring a new PR campaign to change the course of action. Duplicative and triplicate training sessions had to be conducted to remedy the situation, leading to a significant waste of resources and effort.

Lessons Learned:

This situation highlights the importance of effective leadership buy-in when implementing new technology systems. It is unfortunate, but in this case, someones "word" to use the CRM was not good enough. A SLA should have been created. Leaders must align their efforts with their organization's ultimate goals and understand the impact of their actions on the people who work within the organization. It is essential to ensure leaders are fully invested in the implementation process and trained in the system's functionality beforehand. If leaders do not embrace a technological change, their behavior can quickly spread throughout the organization and decrease morale.

In addition, as an organization rolls out a new technological system, it is crucial to ensure that all teams have adequate training and support (at all levels) to make the transition as smooth as possible. Confirming that team leaders understand and are committed to the system's goal ensures a smooth transition. It is also essential to ensure that the new system simplifies workflows and does not create additional headaches. Finally, the information presented should emphasize the benefits of the new system. A clear and concise message regarding a new system could help address anxiety and skepticism and promise the benefits of the new implementation.

Investing time in effective leadership and providing proper education and training is crucial when implementing new technological systems. Additionally, it is essential to communicate often and transparently while ensuring that the new system's goals are understood and embraced. By doing so, an organization can avoid befuddlement, decrease inefficiencies, and fully embrace a new approach to advance its operation.

13

Case Study: Putting All Of Your Eggs In One Basket

Scene:

A company is looking to upgrade its systems, and one of the executives suggests not consolidating all of the technical functionality into fewer, or even one, system. The room is filled with other executives from different departments.

It is amazing when you hear the ways that people come up with to resist change. When you hear of something for the first time, you wonder if it is an oddity, twice a peculiarity, but many times - you know you have a "situation."

This is the story of such a situation. As mentioned, modern CRMs are not cheap; most people think they are expensive. This belief scares people away frequently and, unfortunately before they open their eyes to see the entire picture. This fear does not always manifest itself immediately, nor does not even represent the absolute truth, and is used as an easier-to-understand explanation of their true feelings.

Incident:

One of the executives raises concerns about putting all their eggs in one basket and suggests having more technologies, each with less functionality. Other executives nod in agreement and state similar concerns.

The argument is simple to understand, and an easily persuadable person can brush it off as an acceptable answer. "It is not intelligent for a company to place all its technical functionality within one system. As a wise mentor who's been through it all, I understand the appeal of having more technologies, each with less functionality. It may seem like a logical approach to spread out your technical dependencies and minimize the risk. However, upon closer inspection, it's essential to question the validity of this argument and consider the merits of having less technology, each with more capabilities.

Let's explore some counter-arguments you may encounter when advocating for consolidating your technical functionality within one system:

What if the technology (system) company goes belly up and we have no support?; This concern is reasonable and worth considering. However, it's essential to realize that no matter what vendor you choose, there is always a risk that they may go out of business. By consolidating your technical functionality within one system, you're not increasing the chances of this happening. Rather, you're simplifying your technical footprint and streamlining your operations. To mitigate this risk, you can conduct thorough due diligence on any vendor you're considering. Look at their track record, financial stability, and customer testimonials. Doing so can increase your chances of selecting a reliable and sustainable partner. Additionally, you can have contingency plans to ensure you have access to the necessary documentation, code, and data to transition to an alternative solution.

What if, after overhauling and upgrading all your vital systems, they jack-up the prices and hold you hostage?; While it's always possible that a vendor may manipulate the situation in their favor, it's essential to approach this concern with a proactive mindset. One advantage of consolidating all your technical needs under one roof is having more leverage in the vendor relationship. If there are issues or concerns with the service, you can renegotiate the terms or switch to a different vendor. More technologies with less functionality may make negotiations and the search for alternatives harder. Additionally, having a detailed and clear contract with the vendor can help protect your interests. The service-level agreements (SLAs) and expectations should be clearly stated so everyone understands their responsibilities.

What if you want to migrate to their competition in the future and they make it difficult to do so?; This is a legitimate concern that requires proper planning. However, by having less technology, each with more capabilities, you're simplifying the process of migrating to a different system. When selecting a vendor, it's crucial to assess their data migration capabilities and policies thoroughly. Ensure that they support industry-standard data formats and can facilitate smooth transitions between systems. Doing so can minimize the likelihood of being locked in or facing obstacles when you decide to move on.

Result:

The conversation becomes a debate, and no clear decision is reached. The company's IT team sees the potential benefits of consolidating all technical functionality into one system, but other executives are hesitant due to their concerns.

The result? Reporting processes that sucked your life out like a high-powered vacuum cleaner and decision-making slower than a confused sloth on a leisurely stroll. Do you see my point? This patchwork of "research" in the search for "protection"is fruitless and downright counterproductive.

Lessons Learned:

The fear of putting all your eggs in one basket is a common concern when consolidating technical functionality within one system. However, spreading out technical dependencies and minimizing risks by having more technologies, each with less functionality, can complicate and slow down operations. It can also make negotiations and search for alternatives more difficult.

A detailed and precise contract with the vendor can help protect your interests and ensure that service-level agreements (SLAs) and expectations are clearly stated so everyone understands their responsibilities. Diversifying technical investments and building interfaces and integrations that allow systems to communicate with each other and external solutions reduces dependency on a single vendor and provides a more flexible technological ecosystem.

I will make this somewhat philosophical but stay with me. People who don't want to put their eggs in one basket don't realize the obvious and sometimes painful introspective truth. The basket isn't the technology; the basket is the company itself. We all have one basket. The idea of multiple different technologies webbed together to provide a layer of protection is ludicrous. Here's the ugly truth: The more disparate technologies you try to cobble together, the deeper you're digging yourself into a bureaucratic swamp. You'll need more people, each with highly specialized training, to get the same job done. It's like trying to assemble a team of bearded dragons to change a lightbulb - they may fiercely protect their respective spots on the ceiling, but the bulb's still burnt out.

Oh, and troubleshooting? You'd better have a comfy chair and a strong drink because you're in for a long wait. With each additional system you tack on, the time spent wrestling with these digital gremlins multiplies exponentially. It's a losing battle; you'll be drowning in meaningless error codes before you know it. But here's the kicker - this chaotic mess of a tech stack doesn't even make your data safer. You'll be tearing through more points of potential failure than a bargain-bin document shredder - hardly the impenetrable fortress of data protection folks were so hyped about.

And let's talk about the nightmare that is data weaving. Somehow, you're supposed to stitch all these disparate threads of raw data together to make a quilt of actionable, straightforward information. But instead, it's like trying to tame a hydra with a single hand; the more you wrestle it, the more heads seem to sprout, each more terrible and devious than the last.

What Should Have Been Done Differently:

The executives should have looked at the benefits of consolidating technical functionality within one system and considered the potential risks. Instead of immediately rejecting the idea, they could have done thorough due diligence and looked at the vendor's track record, financial stability, and customer testimonials. This would have increased their chances of selecting a reliable and sustainable partner and mitigated risks associated with such a decision. Moreover, they could have ensured that the vendor supports industry-standard data formats and can facilitate smooth transitions between systems in case of migration in the future. The company could streamline its operations and optimize its strategy by simplifying its technological footprint. Focusing on practicality and common sense instead of trying to cobble together many subsystems can lead to more efficient and effective decision-making.

CONSIDER THIS - THESE COMPANIES, these mighty fortresses of CRM technocracy - they're not the big, bad wolves that some paint them out to be. Quite the contrary! They are your allies, your confidantes, your knights in shining armor. So, let's debunk this shockingly short-sighted myth.

ARE they forcing long and established customers to pay a higher fee? In what universe does that make any business sense? This isn't a battle; it's a symbiotic relationship. CRM companies ensure their survival by keeping you around and helping you succeed.

The logic behind this tall tale is so simplistic that it's almost comical. It's like someone looked at a problem, lost all their critical thinking skills, and thought, "Hey, let's make CRM companies the scapegoat!" Neat package, easy to digest, but devoid of any semblance of intellectual rigor.

Sure, it's got that delicious splash of conspiracy theories. It appeals to the not-so-deep-thinkers desperate for anything to gripe about. It's the perfect cocktail to guzzle down without overthinking about it. But, we are not here for weak sauce whining. We are here for truth, reason, and practicality.

So, here's some advice: don't waste your breath arguing over this. If someone pulls this argument on you, bet your bottom dollar a Trojan horse is hiding amidst their words. There is another issue they're trying to deflect, another pyre they're trying to fan into flame.

Find that issue. Identify it, pull it out, and interrogate it like a spy caught red-handed. Was their grandmother scared by a CRM executive when they were a toddler? Are they scared silly of new technology? Have they gotten too cozy with the older, soon-to-be redundant systems? Or, heaven forbid, are they technophobes in disguise, barking at the moon and mumbling about 'the good old days'?

Time and again, I say this: find the **real** problem. It might be hiding, skulking under layers of excuses – but trust me, it's there. A phobia of being held hostage by CRM providers is nothing more than a smokescreen. Please don't fall for it! Peel away the layers, delve deeper, and demand answers. Only then will you liberate yourself from the perpetual fear of CRM bigwigs and embrace technology in all its liberating, empowering glory.

. . .

I URGE you to cast off the chains of fear and embrace the wisdom that experience has bestowed upon me. Learn from those who have walked this path before you, keep a keen eye on the road ahead, and you'll soon grasp the true crux of the CRM debate.

An educated choice rooted in facts and understanding, not whims or frightening campfire stories, paves the way for strategic growth. So, buckle up, gird your loins, and stride into the brave new world of CRM with enlightened courage. Past the myths and misconceptions, your success story awaits writing. Let's not keep it waiting.

14

Solutions: Out-of-the-Box vs. Build Your Own

The dilemma of selecting a CRM system for your business. It's like being caught between a rock and a hard place, except there are three rocks, and they're all hurling toward you quickly. You've got three options:

- Highly Customized
- Out-of-the-box (OOTB)
- Build From scratch.

Each option has pros and cons, and making the right decision depends on your business's unique needs, resources, and goals. So, sit back, relax, and prepare for a deep dive into the world of CRMs. We'll explore the ins and outs of each approach, and by the end of this book, you'll be equipped to make the best decision for your business.

Customized Solution

THIS OPTION IS like having a personal chef who whips up gourmet meals tailored to your exact tastes and dietary requirements. With a highly customized CRM, you get a system created to meet your specific business needs, making it a powerful tool for streamlining processes and boosting efficiency. However, as with any bespoke service, there's a price to pay. Customized CRM solutions can be expensive and time-consuming to implement, and they often require ongoing maintenance and updates to keep them running smoothly. If you've got the budget and the patience for a tailored solution, this option might be right up your alley.

Pros

Fit To Order To Your Business Needs - A custom CRM can be designed and developed to meet the specific requirements of your business. This means that the system can be tailored to your unique processes, workflows, and data needs, ensuring that it aligns with your business goals and objectives.

Scalability And Flexibility; Custom CRMs should be easily scaled as your business evolves. This means the system can be adapted to accommodate new features, integrations, or changes in your business processes, ensuring it remains relevant and effective over time.

Competitive Advantage; A custom CRM can provide a competitive advantage by offering unique features and capabilities unavailable in OOTB solutions. This can help your business stand out from competitors and provide a superior customer experience.

Integration With Existing Systems; Custom CRMs can be integrated seamlessly with your existing software and systems, ensuring that data flows smoothly between different parts of your business. This can help to increase efficiency and reduce the risk of data silos and inconsistencies.

Cons

Higher Initial Cost; Developing a custom CRM can be costly, especially if you require high customization and configuration. This can impact smaller businesses or those with limited budgets.

Longer Implementation Time; Creating a custom CRM can be a time-consuming process, as it requires the design, development, and testing of the system before it can be implemented. This can delay the realization of the benefits of CRM and may be a disadvantage for businesses that need a solution quickly.

Ongoing Maintenance And Support; A custom CRM may require ongoing maintenance and support from the development team or a third-party provider, which can be costly and time-consuming. This may also make the system more vulnerable to downtime and other issues, as it relies on the expertise of a specific team or provider.

Potential For Feature Overload:;With a custom CRM, there is a risk of developing too many features and capabilities, which can make the system complex and difficult to use. This can lead to low user adoption and reduced effectiveness of the CRM.

Out-Of-The-Box (OOTB)

THIS OPTION IS like walking into a store and grabbing a ready-made meal off the shelf – it's quick, convenient, and gets the job done. OOTB CRMs are generally more affordable and faster to implement than customized solutions, making them a possible option for businesses with limited resources or tight deadlines. However, these systems often have limitations in functionality and flexibility, so they might not be the best fit for companies with unique needs or requirements. If you're looking for a cost-effective solution to get up and running quickly, an OOTB CRM might be the way to go.

Pros

Lower Initial Cost; OOTB CRMs typically have a lower initial cost than custom solutions, as they do not require extensive design and development work. This can make them more accessible for smaller businesses or those with limited budgets.

Faster Implementation; Implementing an OOTB CRM can be quicker than developing a custom solution, as the system is already designed and tested. This can allow businesses to realize the CRM's benefits more quickly.

Standardized Features And Best Practices: OOTB CRMs often include standardized features and best practices that have proven effective in other businesses. This helps ensure that the system is efficient and effective from the outset.

Easier maintenance and support: OOTB CRMs typically come with maintenance and support from the vendor, which can be more straightforward and cost-effective than maintaining a custom solution. This can also help to reduce the risk of downtime and other issues.

Cons

Limited Customization And Flexibility; OOTB CRMs may not offer the same customization and flexibility as custom solutions, making applying the system to your specific business needs and processes difficult.

Potential For Unused Features; With an OOTB CRM, some features and capabilities may not be relevant or valuable for your business. This can lead to unnecessary complexity and make the system more challenging.

Integration Challenges; Integrating an OOTBCRM with your existing software and systems may be more challenging than a custom solution, as the system may not work seamlessly with your specific tools and processes.

Less Competitive Differentiation; Using an OOTB CRM may not provide the same competitive advantage as a custom solution, as the features and capabilities may be similar to those used by your competitors.

Building From Scratch

OOOF: This is a tall order, so I will take it from two distinct perspectives 1) Build from scratch or 2) Build on top of an existing or new CRM.

Build From Scratch; Building from scratch is not a reality for almost everyone. ***Please stop it!*** If you are genuinely asking this question, then actually, the question is, "Can I deal with the pile of crap we have been dealing with for an even longer period (because what you are going to get is a revised version of what you already have)" OR the executives at my company (or worse, you) don't understand what a CRM does. Hint - It isn't a magical spreadsheet from 1995 with 117 columns that have worked wonders until now.

Build On Top (highly customized); Can I customize an accurate and modern CRM to meet my needs?

If you are genuinely thinking about building or continuing, I will give some advice and some potential ammunition to deal with Neanderthal leaders you may be dealing with.

Years ago, there was a void of CRM functionality; there were barren plains where AI-powered data-gathering behemoths now exist. True, customers/investors want frugal companies to drive costs down and make things more economical or gain riches from more significant profits. However, the adage of penny smart and dollar stupid has never been more evident. The irony is that some leaders, and I know a few personally, are so stuck in the past that they never needed such a system, so why do others need such technology? Thank god the Wright brother didn't have this attitude.

If you must, consider that this option best suits businesses with the technical expertise and resources to develop and maintain their own CRM and those with highly specialized requirements that off-the-shelf or customized solutions can't meet.

Allow me to elaborate on the pros of building your own CRM in bullet point format:

Unparalleled Customization; When you build your own

CRM, you have complete control over its features and functionality. You can design it to align precisely with your business's unique needs. By tailoring the CRM to your specific requirements, you can optimize workflows, automate processes, and create an experience that seamlessly integrates with your existing systems. This customization level ensures that your CRM becomes an invaluable asset that enhances your business operations and customer interactions.

Flexibility And Adaptability; One of the big advantages of a custom-built CRM is its ability to adapt as your business grows and evolves. With off-the-shelf solutions, you often find yourself limited by the rigid nature of the system. However, when you have a custom-built CRM, you can easily modify or extend it. This flexibility allows your CRM to keep up with changing requirements, evolving business processes, and emerging technologies. As your business expands, you can add new functionalities, integrate with other systems, and incorporate features that address your evolving needs. Your CRM remains a dynamic tool that grows alongside your business, providing optimal support.

Competitive Advantage; A generic, off-the-shelf CRM may meet basic requirements but lacks the tailored approach to give your business a competitive edge. By building a CRM specifically tailored to your business needs, you have a unique tool that sets you apart from your competitors. A custom-built CRM can provide features and capabilities that your competitors cannot access. It can offer a more efficient way of handling customer interactions, streamline sales processes, and provide detailed analytics that drives informed decision-making. With a finely tuned CRM to your business operations, you can gain a competitive advantage, enhance customer satisfaction, and ultimately increase your market share.

No Vendor Lock-In; Building your CRM in-house eliminates the risk of vendor lock-in, which can often be a significant concern with off-the-shelf solutions. Depending on a specific vendor's platform, you may find yourself locked into their pricing structures, limitations, and future roadmap. By taking the development into your own hands, you have complete autonomy. In-house CRM development allows you to leverage open-source technologies, select your preferred databases, and future-proof your system by staying ahead of industry trends. This independence ensures that you can adapt and scale your CRM according to your business needs without being constrained by a vendor's limitations.

As a seasoned professional in the CRM field, I have witnessed the benefits of building custom CRMs firsthand. By having unparalleled customization, flexibility, and adaptability, a tailored CRM provides a competitive advantage and eliminates vendor lock-in. However, it's crucial to note that building your CRM is a significant undertaking that requires careful planning, technical expertise, and diligent project management. Detailed research, thorough requirements gathering, and proper implementation are crucial to success. With the right approach, building your CRM can deliver exceptional results and provide a powerful tool that empowers your business for years.

Here are a few essential cons to remember:

Significant Time and Effort Required; Developing a custom CRM from scratch is a complex and time-consuming process that demands considerable time and resources. From project management to design, development, testing, and deployment, it's a long and challenging journey that requires meticulous planning and execution. Be prepared to invest heavily in your in-house CRM development in terms of time and money.

Technical Expertise Needed; Building a custom CRM requires a deep understanding of software development, data management, and system integration. Hiring or training specialized staff may necessitate increasing long-term costs and effort. Without the right expertise, you risk improper implementation, which can lead to significant issues down the road.

Maintenance And Updates; A custom-built CRM requires ongoing maintenance to ensure it functions optimally, adding to the long-term costs and effort. While a custom solution can provide precisely what you need at the outset, technology changes and evolving business requirements will force updates to the software, requiring ongoing support. Properly maintaining and updating the system to ensure high availability and performance is critical to the success of your CRM.

Risk Of Over-Complication; Unfortunately, the freedom to create a highly customized solution can make building a CRM more complicated than necessary. Over-engineering the system makes it difficult to use and maintain, leading to future issues and undermining the program's effectiveness. We recommend balancing customization against the system's real-world use on the ground. Some deep documentation can highlight the most critical features of the custom solution, ensuring continued growth and success.

Going Down The Rabbit Hole Is A Genuine Problem; It's all too common to see teams take their customization far too far, creating a confounding array of features that detract from the system's overall usefulness. This problem often arises when non-technical executives dictate the system's functionality without in-depth technical expertise. We recommend involving all stakeholders and extensively prioritizing your system's features to avoid this issue.

When building a custom CRM, the above factors must be carefully considered before embarking on the process. Skimping in one area or another could lead to significant mistakes, causing substantial delays and future financial allocation for remediation. Diligence in planning, execution, and ongoing support will ensure a successful CRM that precisely meets YOUR business needs.

The Dark Side of Choosing the Wrong CRM: Debunking the Waste of Time, Money, and Sanity

LOOK, fellow battle-hardened professionals, we've all heard the siren call of CRMs. They promise improved efficiency, streamlined processes, and driving growth. But don't get swept up in the hype just yet. Before you dive headfirst into the CRM pool without proper research, remember: that not all CRMs are created equal. Choose the wrong CRM, and you'll find yourself drowning in headaches, burning resources, and potentially harming customer relationships. Let's get down to it: no metaphors, analogies, just the cold hard facts.

Get Ready for a World of Hurt: The Side Effects of a Poor CRM Choice

Brace yourselves because choosing the wrong CRM may cause your business a whirlwind of chaos and pain. Here's what to expect if you make a bad choice:

Wasted Resources; We're talking time, money, and for the love of all that is good, your sanity. Investing in a CRM that's too complex, expensive, or time-consuming to maintain could result in pouring resources down the drain faster than a student flushing evidence before a dorm inspection.

Decreased Efficiency; Opting for a system that doesn't align with your business needs means kissing dreams of streamlined processes and increased productivity goodbye. It may just be the exact opposite. It's like trying to change a tire with a toothbrush - impractical, frustrating, and senseless.

Damaged Customer Relationships; Losing sight of the heart of a CRM - managing customer relationships - while opting for bells and whistles you'll never use could hurt your business relationships. Your customers deserve better, and so does your team.

The Brutal Truth: Do Your Due Diligence or Face the Consequences

ALL RIGHT, the cliff notes version didn't scare you off. Good. Let's move on to strategies for mitigating risk and doing your due diligence before taking the CRM plunge.

STEP 1: **Know Your Business**

It's shocking how many business leaders overlook their CRM requirements. That's why it's essential to gather your team and delve deeply into the specifics of your CRM needs. Below are four areas to which you should pay close attention:

- **Pipeline Management:** This feature helps you visualize your sales funnel and understand where each lead is in the sales process. You can track deals, set up reminders, and identify bottlenecks in the pipeline.
- **Lead Scoring:** With this feature, you can evaluate each regarding many things, such as demographics, behaviors, and activities. It enables you to prioritize leads and focus on the ones with the highest potential to convert.
- **Campaign Management:** This feature allows you to create, manage, and track your marketing campaigns across multiple channels. It helps you deliver the right message to your audience at the right time.
- **Reporting and Analysis:** This feature provides insightful reports and analytics that help you analyze the quality of your sales and marketing efforts. Analyzing your data allows you to make intelligent decisions and optimize your strategies for increased ROI.

Remember, identifying and understanding your CRM requirements is critical to selecting a CRM that meets your goals. You're looking for Goldilocks territory: the CRM shouldn't be too complex or simple. It needs to be just right.

STEP 2: **Research, Research, Research**

When selecting the right CRM for your business, it's crucial to approach the decision with careful consideration and thorough research. As a seasoned professional in the CRM field, I've learned that the choices you make at this stage can have a long-lasting impact on your business operations and the happiness of your staff.

Let's break down each point in bullet point format:

- **Now that you know your business inside out, hit the books (books, internet, blogs, colleagues)**: Before diving headfirst into CRM selection, take the time to educate yourself. Conduct thorough research by exploring books, the internet, and industry blogs and engaging with colleagues with CRM implementation experience. Familiarize yourself with different CRM options, best practices, and success stories from other businesses in your industry. Proper research is the foundation for informed decision-making, and neglecting it can lead to scary and costly mistakes.
- **Longevity Matters:** Look for a CRM that has stood the test of time. When selecting a CRM, prioritize longevity and track records. Look for providers with a proven history of reliability and customer commitment. A CRM that has stood the test of time demonstrates its ability to adapt to evolving market trends and technology advancements. Choosing a vendor that has been in the market for a significant period instills confidence. It assures that they will stick around in the long run, continuously supporting and enhancing their CRM solution.
- **Flexibility is Key:** Your business is unique, so find a CRM that can adapt to your specific needs, not a rigid,

one-size-fits-all solution. Avoid the temptation of opting for a one-size-fits-all CRM solution. Your business is unique, with its own set of requirements and processes. Look for a CRM that offers the flexibility to accommodate your specific needs and can be easily tailored to align with your workflows. A CRM that can adapt to your business's unique demands ensures you can optimize your operations and maximize the system's benefits.

- **Integration is More Than a Buzzword:** A CRM should be able to integrate flawlessly with your existing tools and platforms: Integration is a critical aspect to consider when selecting a CRM. A CRM that seamlessly integrates with your in-house tools and platforms enables efficient data flow, streamlines processes, and avoids disjointed operations. Ensure the CRM you choose has robust integration capabilities, allowing for smooth data exchange with your email systems, marketing automation tools, customer support platforms, and any other systems crucial for your business. Failure to prioritize integration can lead to disconnected processes, frustration among your staff, and hindered productivity.

I strongly urge you to approach CRM selection carefully and carefully. Conduct proper research, prioritize longevity, seek flexibility, and ensure seamless integration. Making informed decisions in these areas will set you on the path to CRM success, enhancing your business operations and driving customer satisfaction. Remember, the right CRM solution can be a powerful tool, but the wrong choice can lead to wasted resources, frustration, and hindered growth. Choose wisely and leverage the wisdom of those who journeyed down this path.

STEP 3: **Solicit Opinions**

When evaluating CRM options, going beyond the surface-level

testimonials and digging deeper is essential. You don't want to base your decision on a shiny 5-star rating alone. Real user insights and experiences will provide you with valuable information that goes beyond the marketing pitch.

So, let's break down each point into bullet format:

- **Go beyond the typical customer testimonial:** Talk to actual users of the CRM you're considering and get their unfiltered insights into the software: Customer testimonials can be helpful, but they often represent the best-case scenarios and may not provide a comprehensive picture of the CRM's strengths and weaknesses. Reach out to actual users of the CRM you are considering and engage in conversations to get unfiltered insights into the software. Real user experiences can give you a more accurate understanding of the CRM's performance, usability, and ability to meet your specific business needs.
- **Make Contact**: When connecting with CRM users, target those with similar business models to yours who have also used the CRM software for a comparable duration. This will ensure that the feedback you receive is most relevant to your specific circumstances. Users in a similar industry or with similar operational workflows can offer valuable insights and help you assess the CRM's suitability for your business.
- **Ask pointed questions that elicit critical rather than vague responses**: Don't settle for general praise or broad statements about the CRM. Instead, focus on asking specific questions that prompt users to provide critical answers. Seek how well the CRM has solved problems or helped users achieve their business goals. Ask about the software's ability to adapt to changing needs and workflows. By asking specific questions, you can uncover detailed feedback beyond surface-level impressions.

- **Seek out negative and positive feedback**: Seeking balanced user insights is essential. While positive feedback can indicate areas where the CRM excels, negative feedback can be equally valuable. Knowing what users don't like about the software can help you better understand potential downsides or limitations and determine whether it aligns with your business needs. Pay attention to recurring negative themes and evaluate whether those aspects are critical to your operations.

Engaging with actual users and gathering their unfiltered insights is essential to due diligence. It allows you to make a more informed decision based on real-life experiences. Remember, this isn't Yelp; you want honest answers from users who have experienced the CRM firsthand. By following these steps, you'll gain valuable insights to help you choose the CRM solution that best fits your business.

STEP 4: **Wrangle Those Vendor Demos**

When evaluating potential CRM vendors, it's crucial to roll up your sleeves and dig deep. Don't be afraid to request live demos from each vendor and ask tough questions. Think of it as a Wild West showdown, pitting each CRM against each other to see which emerges victorious.

Let's expand on each point with the benefit of my years of experience in the field, keeping in mind the potential challenges you may encounter:

- **Request live demos from potential vendors - and don't be afraid to ask questions**: A live demo allows you to see the CRM in action and assess its usability, features, and overall fit for your business. Don't hold back during these demos—ask the vendors tough questions. Explore how the CRM addresses your specific pain points and challenges. Consider the ease of use, customization capabilities, and integration options. By scrutinizing each CRM vendor like a seasoned sheriff, you can better evaluate their offerings and make an informed decision.

- **Deployment Matters**: Regarding deployment options, you must consider whether an on-premise or a cloud-based solution best fits your business structure. Both options have pros and cons. On-premise solutions offer greater control and customized security but may require significant upfront investments and ongoing maintenance. On the other hand, cloud-based solutions provide flexibility, scalability, and ease of access, but you'll rely on a third-party provider for security and uptime. Weigh these factors against your business requirements and IT capabilities to determine the most suitable deployment option.
- **Hidden Costs**: The price tag of a CRM system is not the only cost you'll encounter. Beware of hidden fees that can pop up unexpectedly, much like a twisted game of whack-a-mole. When engaging with vendors, ask about any potential additional costs for training, updates, and ongoing support. Some vendors may charge for these services separately, while others may include them in the package. Understanding these hidden costs upfront is crucial to avoid financial surprises later on. Thoroughly review the pricing structure and ensure it aligns with your budget and long-term sustainability.
- **Scalability**: Your CRM needs to keep pace as you and your business grow. Growth is a positive sign, but outgrowing your software can be a costly and time-consuming setback. Consider each vendor's ability to scale with your business when evaluating CRM options. Analyze their track record and testimonials from other companies that have experienced similar growth trajectories. Assess the CRM's capacity to handle increased data volumes, user accounts, and expanded

workflows. Investing in a CRM that can seamlessly grow alongside your business will save you from the headaches of a costly migration down the line.

By following these considerations, you'll be better equipped to navigate the CRM selection process. Request live demos, ask tough questions, and compare vendors like a gunslinger in the Wild West. Discuss deployment options with an understanding of the pros and cons. Beware of hidden costs and ensure scalability aligns with your business growth. With these lessons from my years of experience, you'll be on the right track toward selecting a CRM that propels your business forward. Remember, a well-informed decision today will yield long-lasting benefits for your business.

Test Drive - Vrooom

WHEN CONSIDERING A CRM SOLUTION, evaluating the features and pricing is crucial, and taking the software for a test drive. It's like trying on a pair of shoes before buying them: you want to ensure they fit and don't cause blisters.

Here are three bullet points to remember when taking the "try before you buy" approach:

- **Please take advantage of trial periods and give the CRM a whirl**: Many CRM vendors offer free trial periods to test their software before committing. Please don't skip this step; take advantage of it. This will enable you to assess the software's usability, functionality, and alignment with your business requirements. Test its customization options, integration with other tools, and data import and export capabilities. Inspect the CRM thoroughly, and don't hesitate to ask questions or report issues.
- **Get your team involved, too**: It's essential to ensure that the CRM meets the needs of your entire team, not just the decision-makers. Involve representatives from each department that will interact with the CRM during the trial period. This will give you a more holistic perspective of the software's effectiveness. Discuss the team's workflow requirements, challenges, and pain points. Encourage your team members to provide candid feedback for a more accurate assessment of the CRM's impact.
- **Better to find out now if the system doesn't fit rather than when you're knee-deep in implementation and pot-committed**: Investing in a CRM system can be a significant financial and operational commitment. Ensuring that the software

aligns with your business needs is crucial before investing. Testing the CRM during the trial period can help you identify potential pitfalls early on and save you from making a costly mistake. Better to find out during the trial period if the system doesn't fit rather than when you're fully invested in the implementation.

Remember, testing the CRM during a trial period is a crucial step in the selection process. It helps you better understand the CRM's effectiveness and alignment with your team's workflow requirements. Involve your entire team to get a holistic perspective, and don't hesitate to test the software to ensure it meets your needs thoroughly. Ultimately, a "try before you buy" approach will help you make a better-informed decision and set your business up for success.

Courting the Pushy Partner: Surviving the CRM Vendor's Playbook

PICTURE YOURSELF KNEE-DEEP in the exciting and critical quest for the perfect CRM. You're on a journey to find a tool that goes beyond robotic, predefined platitudes and truly aligns with the unique rhythm of your business. Along this path, you encounter CRM vendors armed to the teeth with glossy demos and impeccably rehearsed sales pitches.

These vendors are well-versed in the art of seduction. They've perfected a well-oiled machine, a veritable assembly line of tactics designed to entice you and make you fall head over heels for their product. But here's the reality, it is your responsibility to see through the smoke screen. You have to interrogate these vendors, scrutinize their claims, and determine if they genuinely have the qualities and capabilities that make them a good fit for your business or are simply trying to shoehorn square pegs into round holes.

In this treacherous landscape, due diligence becomes your guiding star. As an experienced CRM seeker, you know it is crucial to sift through the polished presentations and navigate past the allure of flashy features. This is your opportunity to apply your discerning eye, dig deep into the intricacies of each vendor's offering, and assess whether their solution truly aligns with your specific needs and goals.

Do not be swayed solely by their demonstrations' allure or well-rehearsed spiel. Behind the smoke and mirrors, it is your role to uncover the substance, explore their track record, and assess whether their solution is a harmonious match for the complexity and nuances of your business.

So many, including myself, have fallen into the trap of being dazzled by the surface-level charm of vendors, only to realize later that their promises were empty or that their offerings didn't fit the unique requirements of their organizations. But you armed with the wisdom gained from experience, can rise above this common pitfall.

As you embark on this critical journey, remember the importance of due diligence, meticulous investigation, and insightful questions. Challenge the vendors and press them to demonstrate how their CRM aligns with your strategic objectives, organizational structure, and operational processes. Engage them in detailed conversations about the potential challenges and pitfalls you foresee. Assess their understanding of your unique industry landscape and commitment to ongoing support and partnership.

See beyond the glossy surface - Seek partners who will work alongside you, guiding you toward CRM success. And remember the power is in your hands. Your scrutiny, thorough evaluation, and unwavering commitment to finding the perfect fit for your business will lead you to a CRM solution that meets your needs and harmonizes with your business's rhythm.

So, my seasoned CRM seeker, go forth confidently. Be relentless in uncovering the actual capabilities and value of each vendor's offering. And when you find the one that resonates with your soul, that truly matches the unique heartbeat of your organization, you will know that your diligence has paid off.

May your journey be fruitful and may you find the CRM solution that propels your business to new heights of success.

The Sales Demo - Not a Spectator Sport

WHEN THE TIME comes for CRM vendors to roll out the red carpet, they transform mere sales demos into a spectacle of epic proportions. Picture yourself stepping into the Colosseum, surrounded by the enchanting allure of gladiatorial combat. As the vendors engage in verbal battles, prepare for a clash of wits and tactics.

These vendors are masters of their craft. They will pitch their product as if it were their last day on Earth, pouring on the charm and offering grand promises of how their solution will revolutionize your business. But here's the hard truth: while they attempt to sweep you off your feet and remove any doubts with their persuasive words, you must stay grounded and firmly planted on the ground.

In this exhilarating and dangerous battle, you must remain vigilant. Don't be lured by the flashy presentations and the broad generalizations about how their product will transform your organization. Remember their primary objective is to sell, and it's up to you to see through their captivating performances and determine if their claims have substance.

You must be prepared. Arm yourself with knowledge, research, and a clear understanding of your business requirements. Stepping into the CRM vendor Colosseum without a solid foundation of what you truly need can leave you vulnerable to being taken for a magic carpet ride.

So, how do you keep your feet on the ground amidst the chaos and allure of the vendor battleground? By asking the tough questions. Challenge the vendors to provide specific examples and case studies demonstrating how their product has delivered value to similar organizations.

Don't settle for generalizations. **Push them** to explain in-depth how their solution can address your unique challenges and integrate seamlessly into your existing processes. Demand clear explanations of data migration strategies, system integration capabilities, and ongoing support and training services.

Remember the battlefield is littered with broken promises and shattered dreams of those who did not perform their due diligence. So, be relentless in your pursuit of truth and assurance. Seek references from other customers who have implemented the vendor's CRM and ask for honest feedback about their experiences, and never underestimate the importance of vendor compatibility. When entering the gladiatorial arena, you need more than just a product; you need a partner who understands your business and shares your vision for success. Evaluate the vendor's level of engagement, willingness to collaborate, and commitment to long-term support.

In this relentless battlefield, my seasoned comrade, do not let the enchantment of the vendor performances cloud your judgment. Keep your focus sharp, maintain your skepticism, and trust your ability to see beyond the veils of illusion. By doing so, you can forge fruitful relationships with vendors whose solutions genuinely align with your needs, paving the way to a successful CRM implementation.

Stay wise, stay vigilant, and emerge cautiously from the CRM vendor battlefield. Your journey towards CRM excellence is about finding the right product and the suitable vendor who will stand by your side as a true partner in your quest for customer relationship mastery.

Getting Your Armor On & The Uncomfortable Questions They Fear

Have your interrogation tools ready? You’re not there to be a silent nodding head entertaining the barrage of bullets from their sales cannon. You’re also not there to be sold a pie in the sky. This isn’t a hope trade; it’s a fact trade.

Ask the tough questions. Make them squirm because your business’s victory in the CRM battle depends heavily on their answers.

Here’s a roundup of tough questions CRM vendors might rather face a pack of angry ferrets than answer. But as a seasoned professional, you must ask these questions to ensure you choose the right CRM solution.

Let's delve into each one:

What are the main drawbacks of your CRM?; You can bet this won't be on the sales pitch. They're not going to spell out the shortcomings of their product. It's up to you to dig deeper and uncover the underbelly truth. Ask about the limitations, weaknesses, or areas where the CRM may fall short. Don't settle for vague answers; insist on specific details. You can make an informed decision and mitigate potential risks by dragging the drawbacks into the light.

Have any clients ever canceled, and why?; The juicy details lie within the postmortem of canceled contracts. Learning from others' mistakes can save your business from falling into the same pitfalls. Ask the CRM vendors about any clients who have canceled their subscriptions or terminated their contracts. Understand the reasons behind these cancellations, and analyze whether those reasons align with your business needs and expectations. This knowledge will provide valuable insights into potential red flags or dealbreakers.

How often do you roll out updates, and how disruptive are they?; Updates can introduce new features, interface changes, and workflows that may impact your team's productivity and efficiency. You need to know how often your team's workflow might be disturbed. Ask the CRM vendors about their update frequency and the disruption these updates may cause. Inquire about their change management process and how they support users during these transitions. Assess whether the updated schedule aligns with your business goals and ability to adapt.

What's your data migration support like?; Shifting from another CRM or a manual process to a new system can be complex and challenging. Data migration, in particular, can be a hellhole if not appropriately handled. You don't want to lose crucial data in the transition or end up with orphaned records. Ask the CRM vendors about their data migration support. Inquire about their experience with similar migrations and their methodologies for ensuring a smooth and error-free transition. Assess their data mapping, cleansing, and validation processes to ensure a seamless migration experience.

What level of customization does this CRM offer?; Your business is unique, and a one-size-fits-all CRM may not meet your specific needs. It's crucial to understand how well the CRM offers customization options. Ask the CRM vendors about their system's flexibility and configurability. Inquire about the various customization options, including custom fields, workflows, and reports. Assess whether the CRM can adapt to your business processes and workflows or if you'll be forced to conform to its limitations.

How steep is the learning curve?; Adopting a CRM involves a learning curve, and it's vital to gauge how long your team will need to navigate its twists and turns before they can start driving business growth. Ask the CRM vendors about the learning curve associated with their software. Inquire about the training and onboarding processes they provide to facilitate a smooth transition. Understand how the CRM supports users in getting up to speed and how long it typically takes for teams to become proficient. This information will help you assess the impact on productivity and determine the level of support your group may require.

By approaching CRM vendors armed with these tough questions, you demonstrate your commitment to due diligence. You show them that you're not simply impressed by sales pitches but focused on understanding the true nature of their CRM solution. Remember, it's your responsibility to uncover the answers and ensure that your chosen CRM aligns perfectly with your business needs and goals. Doing so sets your business on a path to success and avoids potential headaches.

Some specific questions for the CRM Commanders on the battlefield

Salesforce

With Salesforce, you're dealing with the 800-pound gorilla of CRMs. But don't let their market dominance intimidate you into mute acceptance.

- **How does Salesforce compare in pricing to other CRM vendors?** – Salesforce isn't exactly penny candy. Their pricing model can get unnecessarily confusing.
- **What exactly does separate licensing mean?**– Salesforce's à la carte pricing structure might get you to rack up hidden costs.

HubSpot

HubSpot has its own resounding horn in the market. Their charm is their unified marketing, sales, and service platform. But don't let that fanfare deter your hard questions.

- **What sort of businesses thrive with HubSpot CRM?** – HubSpot usually caters to smaller companies, so this question can unearth whether their solution fits your business size.
- **Does HubSpot CRM integrate well with non-HubSpot products?** – HubSpot's full functionality shines when you're using their complete software suite.
- Sure thing! Let's step into the interrogation room and prep those questions, guaranteed to make them squirm; got your grilling gloves on? Good. Let's swing away!

Microsoft Dynamics

Microsoft Dynamics is yet another behemoth in the CRM jungle. Their deep integration with other Microsoft products is a key selling point, but probe beyond this face value.

Capable = Yes, very much so. Easy = Nope

- **How well does Microsoft Dynamics play with non-Microsoft products?** – This CRM's wheelhouse is Microsoft software. If you're using other tools, ask if Dynamics will cause friction.
- **Does Microsoft Dynamics require an extended deployment period?** - Dynamics has been known for

its lengthy implementation time. Make sure this fits within your timeline expectations.

SAP CRM

The complexity of implementation and customization may require significant resources and expertise to ensure a successful deployment that aligns with your unique business needs and goals.

- **What's the actual cost?** - SAP's known for holding the "hidden costs" card close to the chest. Implementation costs, operational costs, and user and license management can all add up.
- **Is your tool's interface as user-friendly as you claim it is?** - SAP's had some flak about its user experience. It'll be interesting to see if they've made any advances there.

Oracle Netsuite

Oracle Netsuite may be a formidable force, but its expansive offerings and complex functionality demand scrutiny and evaluation to ensure that it meets specific business needs and avoids costly mistakes down the line.

- **What level of customization can I expect, and at what cost?** - Netsuite isn't exactly the poster child for flexibility. Can it twist to fit your unique workflows, or is it more of a "deal with it" situation?
- **How does your system handle third-party integrations?** - As an all-in-one system, Netsuite may not play well with external tools—time to find out the truth.

Pegasystems

- **What's the learning curve for implementing your CRM?** -With their complex and feature-rich system, this can be a tripwire.
- **What measures are in place to protect my data?** - Security is no joke, and they should be ready to go above and beyond here.

Freshsales

- **What impacts system performance as more functions are added?** - Sure, their CRM looks flawless in demos, but what happens when fully loaded?
- **How flexible is the user count adjustments process?** - Growing or shrinking should be accessible on Freshsales, right? Let's see if it holds.

ZOHO

- **What level of customer support can smaller clients expect?** - Get them in a corner about whether their support teams prioritize more prominent clients.
- **How does your budget-friendly price affect system updates and customer support?** - It's cheap indeed, but will your support calls end up in voicemail oblivion?

Zendesk

- **Do you offer a fully integrated CRM, or should I expect more of a 'ticketing system plus'?** - They're stellar with customer service, but let's see if the CRM capabilities match up.
- **What's the pricing structure for scaling up or down in user numbers?** - They're big on subscriptions, so it's best to ensure you're not signing up for a cash drain.

Agile CRM

- **What steps do you take to reduce users' learning and adaptation curve?** - The name might be Agile, but is the learning process just as nimble?
- **What does CRM migration look like under your guidance?** - Moving house is tough; let's find out if their moving team is good.

Pipedrive

- **Do you offer industry-specific CRM solutions?** - They're big on specializing, but let's dig deep to see how far this extends.
- **How effective is your software for non-sales teams?** - They make a great sales pitch, but how well do they serve your other spearheads?

Zoho

While Zoho might not be the talk of CRM towns like Salesforce or HubSpot, they've got some solid offerings. But no one is free from the hot seat.

- **What sort of improvements has Zoho made regarding the user interface?** – There have been gripes about Zoho's interface. Ask if they've made strides in improving user experience.
- **How well does Zoho's CRM adapt to industry-specific needs?** – Some industry-specific needs may fall through Zoho's CRM gaps.

Wrapping It Up

THE CRM MARKET is a battlefield with vendors as master strategists. They'll try to strike you with dazzling demos and tempting features. But remember, this isn't a game of blind faith. It's about asking the tough questions, digging deep, and getting the right fit for your business.

Take command of the sales demos - make the vendors sweat, question their pitches, poke at the holes. Only then will you stand a strong chance in this face-off. After all, it's your business's growth we're talking about. Just because they come armed with a fancy sword doesn't mean you should let them lop your head off without a fight.

Remember, you're not looking for sweet-sounding answers. You're after the raw, unvarnished truth. They should meet these questions head-on if their claims are as solid as they say. Even a slight deflection can tell you loads about their product if they start dancing around answers.

The Bottom Line: Don't Be That Guy.

You've been warned, but it's worth reiterating: choosing the wrong CRM can wreak havoc on your bottom line, customer relationships, and overall job satisfaction. Adequate due diligence is not just a fancy buzzword combo; it's the key to unlocking a CRM that propels your business forward.

So, before you enter the world of CRM, make damn sure you've done the proper research. Embrace the painstaking process because the alternative is a journey down a nightmare alley nobody wants to take. Your business's success depends on it

15

Setting Expectations Poorly

The Art of CRM Implementation and Beyond

I understand the allure of a shiny new software solution. The world of CRM systems is indeed ever-evolving, filled with exciting possibilities and promises. But before this journey, let's pause and set realistic expectations. As the wise old saying goes, "All that glitters is not gold." Let's take a moment to delve into the art of CRM implementation and what you should reasonably expect during the initial deployment and the following three years.

CRM implementation is not a magic wand that will instantly solve customer relationship challenges. It's a process, a journey that requires careful planning, execution, and continuous improvement. While it's natural to be excited about the possibilities, it's important to approach CRM implementation with realistic expectations and a long-term perspective.

The initial phase of delivering a new CRM is getting the basics right. This includes configuring the system, migrating essential data, setting up user access and permissions, and training your team on its functionalities. It's a critical time to establish a strong foundation, ensuring that the CRM aligns with your organization's goals and processes.

Here's the reality check: CRM implementation is not a one-time event. It's an ongoing journey that continues well beyond the initial deployment. The first three years after deployment are crucial for fine-tuning and optimizing your CRM to serve your organization's unique needs best.

During these early years, you'll likely encounter challenges and roadblocks. It's essential to stay resilient and remain focused on your long-term goals. Expect to invest time and effort in continuous improvement, refining your processes, and adapting to changing customer needs. CRMs are designed to evolve with your organization, so be prepared to embrace change.

TO ENSURE success during this period:

- Focus on critical areas such as user **adoption, data quality, and reporting capabilities**.
- **Encourage and support** your team in fully embracing the CRM and utilizing its features to their fullest potential.
- **Train and retrain** as needed, fostering a culture of continuous learning and improvement.

REPORTING and analytics capabilities are what breathes life into your CRM. Use these features to gain valuable insights into customer trends, performance metrics, and areas for improvement. Leverage these insights to make data-driven decisions that drive growth and enhance the customer experience. Expectations are also frequently misguided due to data quality (which can eminate from beyond the sphere of the CRM) - Keep a close eye on the data being entered, ensuring its accuracy and completeness. Regularly review and clean up data to maintain its integrity and prevent issues down the line. Remember, garbage in, garbage out. Clean data is the lifeblood of effective CRMs.

The art of CRM implementation extends far beyond the initial deployment. It's a journey that requires dedication, perseverance, and a commitment to continuous improvement. Set your expectations accordingly, and understand that success will not happen overnight. Stay focused on the end goal and embrace the opportunities that arise.

Debunking the Silver Bullet Myth

LET'S tackle the silver bullet myth once and for all. Many managers fall into one or more of four traps when introducing CRM. These pitfalls stem from an erroneous assumption that CRM is a standalone piece of software that will handle customer relationships automatically. But that's not the case.

CRM combines customer policies, business processes, and software designed to improve customer satisfaction and financial performance. It's a holistic approach to managing customer relationships, requiring careful planning, execution, and continuous improvement. Establishing a CRM and processes is critical for a company's growth, and it's becoming increasingly expected by existing and new hires alike.

However, there's a common misperception that everything will magically fall into place once a CRM is in place. This myth is a trap that can lead to disaster and is far more believed and practiced daily. This is a huge mistake made by many.

LET ME EXPLAIN.

If your team, leaders, or customers perceive CRM as a silver bullet solution that will automatically solve customer relationship challenges, you're setting yourself up for failure. CRM is not a complete solution; it can often compete with pre-existing techniques operated individually by professionals. The success of a CRM relies on a collaborative effort between technology, people, and processes. It is widespread for customers and leaders to think like this "Well, it costs $$$$. Therefore, it must fix everything on its own." What a pitiful and sorry excuse for a system being anemic or flat-out failing. But as I said, it happens a lot.

It's essential to recognize and communicate (from way before deployment) that CRM is not a magical key that will open doors to customer loyalty and financial success automatically. It's a tool that can improve user results and procedures when combined with sound policies and processes.

For executives and leaders, it's critical to provide training and support throughout the implementation process and beyond. Creating a continuous learning culture would be best, encouraging your team(s) to embrace the new system actively. CRM is a journey, not a destination, requiring ongoing attention and investment.

For customers, a CRM can be a powerful tool that improves their experience with your company. However, it's essential to communicate, transparently and upfront, how your CRM works, respects their privacy, and how customers can engage with it easily. This way, you can manage expectations and gain their trust, leading to better customer satisfaction, loyalty, and financial success.

There is no silver bullet or even an easy solution when it comes to CRM. It's a complex ecosystem that requires careful planning, execution, and continuous improvement. With the right approach and mindset, however, you can leverage CRM to drive customer satisfaction, employee productivity, and financial performance.

Finding the Sweet Spot

CUSTOMERS, in all their wonderful variations, will have certain expectations that you, as the provider, may not have been planning to deliver. And on the flip side, if you're the customer, you may find yourself expecting something that won't be delivered. Hold on a minute, don't jump to conclusions, and assume foul play. This misalignment often results from preconceptions and misconceptions that haven't been laid out on the "table" consistently.

How do we avoid this potential pitfall? The answer lies in establishing a clear understanding of the current state of your pre-CRM process. This can be achieved by delving into the nitty-gritty details through a thorough "day in the life" analysis of the users. Think of it as going undercover, immersing yourself in the daily activities, workflows, and pain points of those using the CRM. Through this deep examination, you'll be able to uncover the hidden challenges and identify the gaps that need to be addressed.

Once you've completed this analysis, holding a debriefing session is crucial to discuss and share the findings. This is where both sides come together to align expectations and bridge the gap between what's currently in place and what's needed for the CRM to deliver the required and expected improvements. It's a platform for open and honest communication where all parties can discuss their perspectives, clarify any misunderstandings, and outline the scope of the solution.

Communication is the key. Don't assume everyone is on the same page or that expectations are crystal clear. Keep the lines of communication open throughout the process, and don't be afraid to ask those tough questions. Doing so uncovers any hidden expectations or misconceptions that can cause trouble.

Regarding CRM implementations, finding a spot between what's realistically achievable and expected is essential. This requires a collaborative effort between all stakeholders, with a shared understanding of the current state and a clear vision of the desired improvements. Don't shy away from rolling up your sleeves and getting into the details. Through this careful examination and open communication, you'll be able to ensure that the deliverable aligns as closely as possible with the requirements and expectations.

Let's talk about setting expectations. It's a critical component of any project, especially regarding CRM implementations. Without a clear understanding of what success looks like, it's challenging to create a roadmap or measure your progress. That's why using something relatable, tried, and true at your organization is advisable.

It is imperative for setting expectations that you have a clear roadmap that sets out the journey, indicating what milestones will be achieved and when and how long the whole process will take to complete. This is a critical point. One of the biggest frustrations for people implementing a CRM is not knowing how long it will take and when they'll see results. This can lead to an experience that feels like it's never-ending, and all people involved know where they (and their solutions) are in the process.

A timeframe for many months and years is a great way to set expectations. Why is that, you ask? Well, a timeline of three years, for example, is long enough for you to get traction and make sustainable progress but not so long that it feels like you're not making any headway. It's also a timeframe relatable to many businesses, regardless of size or industry. By setting a clear timeline, you can drive progress and set leaders' minds at ease, showing them that CRM success isn't a "snap of the fingers" win.

When setting expectations, it's crucial to be realistic and honest. You can't promise the world if it's impossible, as this will only lead to frustration. Instead, create a tangible roadmap that can be measured and adjusted. And don't forget to communicate this roadmap clearly to all stakeholders, from the top down to the front lines. Remember to be realistic and honest, providing a tangible roadmap that can be communicated clearly to all stakeholders. With the right expectations, you can measure progress and achieve CRM success.

Review the following paragraphs and keep in mind that these steps are just rough guidelines, and you'll need to customize them to fit your style and what you believe will resonate with the stakeholders you're aiming to keep happy.

There is no one-size-fits-all solution when it comes to CRM. Every organization is unique, and what works for one may not work for another. So, don't be afraid to tweak these ideas and make them your own

The journey is all about discovery and growth as we navigate the post-implementation phase of your new CRM. To ensure success, we need to keep a realistic mindset and embrace the spirit of continuous improvement. So, let's dive into the key milestones and expectations that will guide us on this exciting adventure:

YEAR 1: **The Honeymoon Phase**

In this first year, our focus should revolve around these three crucial points:

Smooth Transition; We want to ensure all users' seamless adoption of the CRM. This means providing comprehensive training and support to help employees become comfortable and confident with the new software. You need to be there every step of the way, ensuring everyone embraces this change without any roadblocks.

Measure Success; To gauge the effectiveness of the CRM implementation, we must establish clear metrics and benchmarks. These will help us track progress and determine if we're meeting our goals. By defining success and how we'll measure it, we can ensure that our efforts align with the desired outcomes.

Lay The Foundations; Year one is also the time to establish the groundwork for future growth. We'll refine our processes, adapt CRM policies and strategies to align better with the CRM's capabilities, and identify any areas for improvement. Setting a solid foundation early on will pay dividends in the long run.

. . .

YEAR 2: **The Reality Check**

Ah, the second year—the reality check phase. Here's where we get down to business and fine-tune our approach:

Leverage Insights; Now that we have a year of data and insights generated by the CRM, it's time to analyze and harness this wealth of information. By identifying areas for improvement and optimization, we can make data-driven decisions that enhance our operational efficiency and customer interactions.

Refine and Adjust; CRM is not a static endeavor. We must continuously refine and adjust our strategies, policies, and procedures to fully leverage our CRM's capabilities. This means staying agile, adapting as needed, and **aligning** our practices with industry best practices.

Monitor and Measure; The success of your CRM should never be taken for granted. Regularly monitoring and measuring its impact will ensure that it continues to deliver value and meet the evolving needs of your business. We can quickly identify deviations from our goals and make necessary course corrections by staying vigilant.

YEAR 3: **The Road to CRM Maturity**

Welcome to the final leg of our CRM Odyssey—the road to maturity:

Enhance Effectiveness; By consolidating the learnings and best practices gathered over the previous two years, we can further enhance the effectiveness of your CRM. This includes refining processes, optimizing workflows, and streamlining operations to maximize efficiency.

Embrace Innovation; It's time to leverage your CRM's capabilities to drive new initiatives and explore untapped markets. Let's identify opportunities for innovation, such as implementing advanced analytics or launching new customer engagement programs. By embracing innovation, we can gain a competitive edge and unlock new avenues for growth.

Nurture Continuous Improvement; We must establish a culture of continuous improvement, ensuring that the CRM remains a cornerstone of your company's long-term CRM strategy. Encourage your teams to provide feedback, share ideas, and embrace change as we continually refine and optimize practices.

WHEN IMPLEMENTING A CRM IMPROVEMENT PLAN, it's essential to emphasize that this is a long-term endeavor. Continually reminding others it is a multi-year plan will set the right expectations and ensure everyone understands the journey ahead. Additionally, it's crucial to emphasize that this plan will be followed by additional multi-year plans to ensure continual growth and keep up with evolving customer needs and market trends.

Creating a culture of participation is critical to the success of your CRM efforts. Each person in your organization should feel a sense of ownership and responsibility for the success of the CRM initiative. From the top executives to the front-line staff, everyone should understand the importance of monitoring seemingly minor issues and addressing them promptly via published and known paths of escalation.

Encourage individuals to be vigilant and proactive in identifying any challenges or opportunities that arise during the implementation of the CRM plan. This could be anything from data entry errors to process bottlenecks or customer feedback. Everyone should feel empowered to bring these concerns to the attention of the larger group.

By fostering an environment where small issues are brought to the collective attention, you can ensure that no problem goes unnoticed or unaddressed. This allows the larger group to assess and plan accordingly, making informed decisions to mitigate risks, seize opportunities, and continuously improve the CRM effort.

Remember, Rome wasn't built in a day, and the same goes for CRM success. By acknowledging that it's a multi-year plan and enlisting the participation of every individual in your organization, you'll be laying the groundwork for long-term growth and sustainable success. Stay committed, remain vigilant, and tackle challenges as they arise. With a collaborative approach and a focus on continual improvement, your CRM effort will yield significant benefits over time.

The CRM Conundrum: Left Unsatisfied and Stuck

PICTURE THIS: it's been two to three years since your company implemented a shiny new CRM system. You had high hopes for it, envisioning seamless customer interactions, streamlined processes, and a clear return on investment (ROI). But instead, you find yourself scratching your head, wondering why the CRM isn't living up to its promise. You're not alone. Many companies experience dissatisfaction with their CRMs post-implementation, often due to a lack of clear ROI, unimproved processes, poor adoption, and an overall sense of unease about the system's benefits.

Issue 1: The Elusive ROI,

One of the primary reasons companies find themselves dissatisfied with their CRMs, is the difficulty in determining a clear and easily understood ROI. After investing time, money, and resources into implementing a CRM, expecting a tangible ROI is natural. However, measuring the ROI of a CRM can be complex and nuanced, as the benefits often extend beyond simple financial metrics.

Likely Reason For Unmet Expectations: Measuring CRM ROI often stems from unrealistic expectations and a lack of clarity around what constitutes success. Companies may expect immediate, easily quantifiable results when in reality, the benefits of a CRM may take time to manifest and can be influenced by numerous external factors.

Potential Solution: To better understand the ROI of your CRM, consider adopting a more holistic approach to measurement. This could include assessing quantitative factors (such as revenue growth, cost savings, and efficiency improvements) and qualitative factors (such as customer satisfaction, employee engagement, and collaboration). By taking a broader view of the CRM's impact, you may be better equipped to appreciate its true value and potential.

. . .

ISSUE 2: **Stagnant Processes**

Another common source of dissatisfaction with CRMs is the perception that they have not improved processes that may or may not have been reviewed and tweaked before transferring into the CRM. This can lead to upset customers and disillusionment with the system's capabilities.

Likely Reason For Unmet Expectations: The failure to improve processes within a CRM may be because the methods sucked, to begin with, and not enough work was done to improve them before being transferred into your CRM. However, the true potential of a CRM can only be realized when it is used in conjunction with well-designed, efficient processes.

Potential Solution: To address this issue, consider thoroughly reviewing your existing processes and identifying areas for improvement before transferring them into the CRM. This could involve streamlining workflows, eliminating bottlenecks, and automating manual tasks. Once the CRM is in place, continue to monitor and refine these processes to ensure they remain optimized and aligned with your company's goals and objectives.

ISSUE 3: **Poor Adoption**

A third factor contributing to CRM dissatisfaction is poor adoption among employees. If your team is not fully utilizing the CRM, it's unlikely to deliver the expected benefits. This can result in frustration and disappointment with the system's performance.

Likely Reason For Unmet Expectations: Poor adoption of a CRM may be due to a lack of communication and understanding among employees. If your team is not aware of the benefits of CRM and how it can improve their daily tasks, they may be resistant to change and hesitant to embrace the new technology.

Potential Solution: To boost CRM adoption, focus on providing comprehensive training and support to your employees. Ensure they understand the benefits of CRM and how it can make their jobs more accessible and efficient. Additionally, consider incorporating the CRM into your company's culture and performance metrics, rewarding employees for their engagement with the system, and encouraging its use across the organization.

ISSUE 4: **Overall Uneasiness**

Finally, many companies experience a general sense of uneasiness with their CRMs, unsure of what the system is doing better for the company. This can lead to a growing lack of confidence in the CRM and a reluctance to invest further resources into its development and optimization.

Likely Reason For Unmet Expectations: A CRM's overall uneasiness may be due to a lack of communication and clarity around the system's purpose and capabilities. Companies may have unrealistic expectations about what CRM can achieve or not fully understand its potential benefits. By addressing these misconceptions and fostering a culture of collaboration, companies can help alleviate this uneasiness and unlock the true potential of their CRMs.

Potential Solution: To alleviate this uneasiness, maintain open communication lines between all stakeholders in the CRM. Review the system's performance regularly, discuss issues or concerns, and celebrate successes and milestones. By operating a culture of transparency and collaboration, trust will grow in the CRM and ensure it remains a valuable asset to your company.

CRM DISSATISFACTION IS a common challenge many companies face, but it doesn't have to be a permanent state of affairs. By addressing the issues of ROI, process improvement, adoption, and overall uneasiness, you can transform your CRM from a source of frustration to a powerful tool for growth and success. So, embrace the CRM journey with a healthy dose of realism and a touch of wit, and prepare to reap the rewards of a well-implemented and well-utilized CRM.

16

Case Study: Not Being Realistic

Scene:

Imagine the ramp-up months leading to a massive rollout of modern CRM systems. Amidst the preparations, a group of internal onlookers discovered a shocking truth about the outsourced CRM project discovery and deployment team—**they were inept.**

Initially, the onlookers saw this as an opportunity to educate the project leaders further and surpass their definition of success. However, they soon realized that the leaders had different motivations. Their primary concern was to avoid being schooled or making disruptive changes that could rock their perception of a smoothly sailing ship. They sought peace of mind, recognition for their decision to implement CRMs, and the assurance that they had provided a valuable toolset for the global staff. There was also a desire to get the work done and get paid. Unfortunately, these subtle nuances went unnoticed, leading to disastrous consequences.

Incident:

The incident unfolded as the ineptitude of the outsourced CRM project team became apparent. Their lack of expertise and understanding of the organization's goals and unique requirements became evident as the ramp-up months progressed. What initially seemed like an opportunity to educate and surpass expectations became an uphill battle. The leaders, blinded by their desire for recognition and a frictionless deployment, resisted any suggestion of change or additional guidance.

Result:

The result of neglecting these subtleties and misjudging the true motivations of the leaders was catastrophic. The consequences of the inept CRM project deployment team's actions and the resistance to education and adaptation became apparent in the following ways:

Underutilized CRMs; The global staff did not effectively utilize the CRMs due to insufficient training and guidance. The leaders' unwillingness to learn and adapt resulted in missed opportunities to leverage the full potential of the CRMs.

Failed Rollout; The large-scale rollout of modern CRMs, which held the promise of streamlining operations and enhancing productivity, fell flat. The lack of expertise and resistance to change led to a subpar deployment that did not fulfill its intended purpose.

Damaged Trust and Morale; The failure of the CRM implementation eroded trust among the staff. Employees became disillusioned, questioning the leadership's decision-making and commitment to supporting their teams with practical tools and processes.

Lessons Learned:

This incident highlighted several critical lessons organizations should consider when deploying CRMs:

Proper Due Diligence; Thoroughly vet and assess any

external teams or vendors involved in the CRM implementation. Evaluate their expertise, alignment with your organization's goals, and ability to adapt to unique requirements.

Align Expectations; Communicate with project leaders and stakeholders to ensure a shared understanding of what success looks like. Identify and incorporate their motivations into the project plan while advocating for the necessary education and flexibility to achieve optimal outcomes.

Emphasize Change Management:;Implement a comprehensive change management strategy that accounts for the resistance to change that may arise from leaders focused on peace of mind rather than growth. Addressing this resistance through education, communication, and support is vital for successful adoption.

What Should Have Been Done Differently?:

To avoid the consequences faced in this incident, the organization could have taken a different approach:

Hire A Much More Competent and Caring Implementor With Deeper Engagement; The internal onlookers should have engaged more closely with them instead of assuming that the outsourced CRM project team would meet their expectations. By actively participating in the discovery and deployment phases, they could have identified ineptitude early on and provided additional guidance.

Managing Expectations; It was crucial to have frank discussions with the leaders regarding their motivations and desires for the CRM implementation. The organization could have fostered a more collaborative and receptive environment by addressing their concerns and clearly defining the expected benefits.

Educating Leadership; Recognizing the resistance to change, efforts should have been made to educate and communicate the value of CRM beyond just providing a toolset. Demonstrating how CRM adoption can improve efficiency, revenue growth, and better customer relationships could have helped align the leaders' motivations with the project's goals.

. . .

ORGANIZATIONS CAN ENHANCE their chances of successful CRM implementation by being vigilant, aligning expectations, embracing change management, and reaping the benefits of improved sales processes and customer relationships.

17

Repeating The Same Mistakes That Old-World Thinking Induces

Old-world Thinking in the realm of CRM. It's an obstacle that many businesses face as they try to navigate the ever-changing landscape of CRM. CRM has come a long way since its early days, but unfortunately, some folks are still stuck in the past.

The problem lies in the outdated definition of CRM that still lingers in many people's minds. They view it as a simple contact management system, almost like a big corporate Rolodex in the sky. And let me tell you, this misconception can cause quite a sharing problem.

When people believe that CRM is just about collecting their information, contacts, and sales tricks, they hesitate to engage with the system entirely. They worry that their hard-earned insights and personal connections will be snatched away, lost in some data-hoovering abyss. This fear can hinder collaboration and the effectiveness of the CRM effort and ultimately impede business growth.

What are these old-world ways of thinking that act as a hindrance? Well, it's the resistance to embracing the true potential of CRM. Example manifestations and reasons:

- They didn't need these "new fangled tools," so why should anyone else?
- The fear of change and the desire to cling to familiar methods can prevent experienced and older professionals from fully trusting and accepting the truth about CRM's capabilities.
- Perceived protection of their lifelong developed intellectual property.

LET'S be contemporary - New ideas that may be hard for some to embrace are the very ones that can revolutionize their business. It's about recognizing that CRM has evolved into something more powerful. It's not just about managing contacts anymore; it's about harnessing the power of data, automation, and customer engagement.

These new ideas are about using CRM as an invaluable tool for analyzing customer behavior, streamlining workflows, and personalizing interactions. They're about leveraging technology to improve customer service, increase efficiency, and drive business growth. It's a paradigm shift that requires an open mind and a willingness to let go of old habits.

So, how can we bridge the gap between old-world thinking and new CRM possibilities? Education and communication are essential. We need to help experienced and older professionals understand the true potential of CRM, debunk the myths, and show them how it can enhance their work, not replace it.

We can share success stories and case studies of businesses that have embraced the modern CRM approach and reaped substantial rewards. We can provide training and support to help these professionals adapt to the new tools and techniques. And perhaps most importantly, we can create a supportive environment where experimentation and learning from mistakes are encouraged.

Educate, communicate, dispel misconceptions, and show experienced and older professionals the incredible possibilities with an open mind and a willingness to adapt and adopt.

18

Integrating CRM With Existing Business Processes

CRM system integration into your existing workflows can be a hassle if you're not careful. But they are vital for many reasons, including:

- Helping you understand your customers.
- Enabling you to respond more quickly and effectively to customer inquiries.
- Allowing you to track important metrics such as revenue, sales funnel stages, and pipeline data.

THIS CAN BE a complex and challenging process. However, successful integration is crucial for minimizing disruptions and maximizing the benefits of the CRM. One of the critical aspects of CRM integration is modifying existing workflows and processes to incorporate the new technology. Exploring the key steps and considerations for modifying workflows and processes during CRM integration and the potential problems that may arise if these modifications are not made.

. . .

ANALYZE AND IDENTIFY NECESSARY CHANGES; The first step in modifying workflows and processes to incorporate a CRM is to analyze the existing workflows and identify the necessary changes. This involves understanding how the CRM will impact various departments, such as sales, marketing, customer service, and IT, and determining which processes need to be modified or redesigned to accommodate the new technology. Organizations can ensure a smoother integration process and minimize disruptions by identifying the necessary changes.

Create a Change Management Plan; Organizations should create a comprehensive change management plan once the necessary changes have been identified. This plan should outline the steps and timelines for implementing the changes and any potential risks and contingencies. Organizations can better manage the transition to the new CRM by developing a detailed change management plan and minimizing the risk of disruptions and system failure.

Communicate and Involve Employees; Modifying workflows and processes to incorporate a CRM requires the involvement and support of employees at all levels of the organization. It is essential to communicate the reasons behind the changes and the benefits they can expect from the new system. Involving employees in the change process and ensuring they understand the latest workflows and processes can help minimize resistance and accelerate user adoption.

Provide Training and Support; Implementing new workflows and processes often requires employees to learn new skills and adapt to new working methods. Organizations should provide the necessary training and support to help employees transition to the new CRM and its associated workflows and processes. This may involve offering workshops, online training resources, or one-on-one coaching to ensure employees are well-prepared for the changes.

Monitor and Adjust; After implementing the new workflows and processes, monitoring their performance and making necessary adjustments is essential. This may involve refining workflows, addressing technical issues, or providing additional training and support to employees. Organizations can remain aligned with the CRM and deliver the desired results by continuously monitoring and adjusting workflows and processes.

CAUTION: If workflows and processes are not modified to incorporate the new CRM technology, several problems will be encountered.

THE most commonly realized problems are:

Inefficiencies and Bottlenecks; Failing to modify workflows and processes to accommodate the CRM can result in inefficiencies and bottlenecks, as employees struggle to adapt to the new technology and continue using outdated processes. This can create decreased productivity and increased frustration among employees.

Reduced CRM Benefits; If workflows and processes are not modified to align with the CRM, organizations may not fully benefit from the system's features and capabilities. This can result in a lower return on investment and a failure to achieve the desired business outcomes.

Employee Resistance; Employees may resist using the CRM if workflows and processes are not modified to accommodate the new technology. This resistance can lead to a lack of user adoption, resulting in system failure.

Data Inconsistency and Inaccuracy; Failing to modify workflows and processes to align with the CRM can result in inconsistent and inaccurate data, as employees may continue using outdated data entry and management methods. This can negatively impact decision-making and overall business performance.

Increased Risk of System Failure; If workflows and processes are not modified to incorporate the CRM, there is a higher risk of system failure due to inefficiencies, bottlenecks, and employee resistance. This can facilitate wasted resources, lost opportunities, and damage to the organization's bottom line.

Organizations can ensure a smooth and seamless CRM integration by analyzing and identifying necessary changes, creating a change management plan, communicating with and involving employees, providing training and support, and monitoring and adjusting workflows and processes. Failure to modify workflows and procedures to accommodate the CRM can result in numerous challenges that can hinder the effectiveness of the CRM and negatively impact the organization's bottom line.

19

Making Organizational Changes

Implementing CRM Before Making Organizational Changes

Preparing the employees and your organizational structure is the first step to successful CRM deployment. Think of it this way. You can't paint a house without a primer. It won't last as long, and the cracks only start appearing from the first coat.

Installing the software before establishing a customer-focused company/department/division is the most hazardous stumbling block to implementing CRM technologies. If a company wishes to build stronger relationships with its most profitable customers, it must first overhaul critical customer-facing business operations, such as customer service and order processing. To this point, remember that your customer as a CRM operations professional is practically everyone - this includes external customers, internal users, leadership who receive reports, other technical systems that give/get information, and other people in other departments that depend on the systems supplied data as being rock solid.

Having a customer strategy and calling it a day is not enough. A CRM deployment succeeds only after the company and its processes —job requirements, performance metrics, compensation schemes, training courses, and so on—have been reorganized to meet consumers' needs better.

Your employees are key stakeholders in how the company functions. While you may be in charge of strategy and big business ideas, the staff you hire is crucial to running your company. If you want them to be on board with what is quite a transition, you need to prepare them well in advance.

Assessing current departmental, commodity, and regional systems is also critical. Executives also overlook improving internal frameworks and procedures before engaging in CRM technology. A dangerous mistake is assuming that CRM only impacts customer-facing systems.

CRMs are well-known for being able to monitor user activity. Tasks completed, maintained records, emails received, and measures evaluating sales success are just a few examples. Some salespeople are concerned that the CRM will be used against them if they fail to fulfill their quotas. They do not intend to sabotage data security. Still, if the CRM is considered challenging, they will only reluctantly enter data as necessary and almost certainly not double-check it for validity.

Spreadsheets are easy, accessible, free, and well-known. When monitoring leads, sales reps who have worked as consultants or at several companies also have ingrained behaviors. They could do well in this situation.

When deploying CRM or any other technology, the first thing to do is better let the people who will be working with it know what they are working with.

20

Case Study: Initial Design - Too Many People, Q&A, And Siloed Input

Scene:

Picture a bustling project team room filled with individuals from different departments and teams. Each person brings their unique perspective, goals, and objectives to the table, creating a sense of excitement and eagerness to contribute. However, amidst this buzzing energy, there is a palpable feeling of chaos and a distinct lack of direction.

In these situations, it's crucial to recognize that any CRM implementation's success hinges on aligning and harmonizing the efforts of diverse teams.

While enthusiasm is commendable, the project can quickly veer off course (foreshadowing) without a clear direction, leading to unmet expectations and wasted resources.

Incident:

Stakeholders advocated their goals passionately, and each was convinced that their perspective was the key to success. This clash of priorities leads to frequent disagreements and delays in decision-making, impeding progress and threatening the project's timeline.

During these situations, I have witnessed how adamant stakeholders can be about their goals, which often results in struggles to reach a consensus with others. Each stakeholder argues their case, convinced that their objectives and strategies are best for the organization. While reflective of a desire to make meaningful contributions to the project's success, these impassioned debates can create an impasse that halts progress and diminishes the chances of achieving desired outcomes.

In my years of experience leading CRM implementation projects, I have encountered numerous situations where a **collaborative environment can, paradoxically, hinder the progress of a project.**

As discussions unfold within this collaborative environment, conflicting priorities and differing opinions inevitably emerge.

Result:

It's a common pitfall that can quickly derail even the most well-intentioned initiatives. As deadlines were missed and key deliverables were constantly delayed, progress toward the finish line became painfully sluggish. The lack of coordination and alignment significantly hampered the project's ability to move forward effectively, resulting in frustration that started to build among team members.

One of the biggest lessons I've learned from this experience is the importance of having clear goals and a shared vision for the project from the beginning. It may sound obvious, but it's surprising how often this crucial step is overlooked. By clearly defining what needs to be achieved and ensuring that all stakeholders are on the same page, teams can work towards a common objective, fostering collaboration and minimizing self-interest.

Another critical factor in avoiding these pitfalls is fostering open communication and regular checkpoints. Establishing a culture where team members feel comfortable speaking up and providing updates can significantly reduce the chances of misalignment and costly delays. By holding regular meetings and encouraging cross-team collaboration, potential conflicts or misunderstandings can be identified and resolved early on, preventing them from escalating into significant roadblocks.

While it may seem tempting to dive right into a project without taking the time to plan and align, the consequences can be severe. The project may stumble and meander, wasting valuable time and resources. So, as you embark on your projects, I cannot stress enough the importance of due diligence and meticulous planning. Take the time to establish clear goals, foster open communication, and continuously evaluate progress. By doing so, you'll be better positioned to overcome obstacles and guide your team toward success.

Lessons Learned:

To navigate such challenges, it is crucial to establish a structured decision-making process that allows for the open exchange of ideas while also maintaining focus and progress. Designating an apparent decision-making authority, such as a project lead or an executive sponsor, helps streamline discussions and resolve conflicts when consensus proves elusive. This individual, leveraging their expertise and authority, can help guide the team toward productive decision-making and ensure the project remains on track.

In addition, actively fostering an environment of respectful communication and collaboration is paramount. Encouraging stakeholders to listen to one another, empathize with differing perspectives, and find common ground can help create synergy and enable progress even amidst conflicting goals. Facilitating discussions that address concerns, explore options, and seek compromise allows the team to move forward, pooling their collective knowledge and experience toward a unified vision.

I cannot stress enough the importance of **managing stakeholder expectations from the outset**. It is crucial to involve stakeholders in the early stages of the project, ensuring their input is heard and integrated into the project plan. This involvement helps foster a sense of ownership and shared responsibility, reducing the likelihood of conflicting goals arising later in the project.

In conclusion, I have learned through experience that a collaborative environment, while beneficial, can create challenges when too many stakeholders with conflicting goals are involved in CRM implementation projects. To overcome these obstacles, it is essential to establish a structured decision-making process, designate an apparent decision-making authority, foster open communication and collaboration, and actively manage stakeholder expectations. By doing so, organizations can pave the way toward successful CRM implementation, leveraging the collective expertise of stakeholders and overcoming obstacles that hinder progress.

Establish clear project objectives; Defining and communicating the project's goals and objectives from the outset is crucial. This ensures that all team members understand what needs to be achieved and helps align their goals with the bigger picture.

Encourage collaboration and open communication; Foster an environment where team members feel comfortable openly discussing their goals, concerns, and ideas. Encourage active listening and create channels for effective communication to facilitate collaboration and prevent misunderstandings.

Embrace Agile methodology; Agile methodologies like Scrum or Kanban promote iterative and flexible project management approaches. By breaking the project into smaller, manageable tasks and implementing regular review and feedback cycles, teams can adapt to changing circumstances and deliver value incrementally.

What Should Have Been Done Differently:

To avoid finding yourself in a situation where too many people with too many goals hinder project completion, specific steps could have been taken:

CREATE A DEDICATED PROJECT LEADERSHIP TEAM; Establish a small, cross-functional team responsible for guiding the project, setting priorities, and making critical decisions. This team should represent different stakeholders and have the authority to align goals and resolve conflicts.

Perform stakeholder analysis and engagement; Identify key stakeholders and their objectives early in the project. Engage with them to understand their needs, manage their expectations, and ensure their goals align with the project's objectives.

Implement Agile project management practices; Agile methodologies provide benefits such as increased adaptability, faster feedback loops, and regular course correction. These practices, including sprint-based planning and regular retrospectives, would enable better collaboration and help manage conflicting goals.

BY TAKING these proactive measures and implementing Agile methodology, teams can better manage multiple goals, facilitate effective coordination, and increase the project's chances of reaching the finish line successfully.

21

Case Study: You're Not The Owner, I Am.

Scene:

Picture a bustling corporate office filled with ambitious individuals striving for success. In this scenario, a group of seemingly honest people becomes caught up in self-serving events that lead to a series of self-serving decisions. Although not necessarily evil, these individuals are unaware of the harm they are causing. They have an overwhelming desire to take ownership of everything they or their divisions touch, regardless of whether they are the sole users of that information. They exhibit a complete lack of concern, appreciation, and openness to other ideas.

Unfortunately, these individuals are more common than most people would like to believe and are often characterized by their stubbornness and bullying tendencies.

Incident:

Within this corporate setting, we find several adult children who work for the same company. Each of them becomes so fixated on their own "fiefdoms" – their respective departments or areas of responsibility – that they forget about the overarching goals of the organization and the importance of CRM systems. Their primary focus becomes growing their sphere of influence and their teams, despite being part of a larger organization that relies on jointly owned knowledge and processes. Instead of working towards the collective well-being of the organization, they make decisions solely based on their own self-interest.

To further their cause, these individuals resort to exaggeration and primarily false claims to present their areas as performing better or needing more attention than other aspects of the company. They even go as far as cutting from the budget or manipulating strategy reports to serve their own interests and gain control over their territories.

Result:

Regrettably, the consequences of these self-serving decisions can be detrimental to the company in the long run. As each individual focuses solely on their own interests and neglects the bigger picture, the organization loses sight of its primary goals and the collective well-being of its customers. This one-sided approach leads to losing time, money, and trust from customers and stakeholders alike. Ultimately, the company's success and future growth are jeopardized.

Let me emphasize the danger of self-serving decisions. Although it may seem advantageous in the short term, focusing solely on one's interests can have disastrous consequences for the company in the long run. This type of behavior leads to a reduction in the collective well-being of customers, stakeholders, and the organization as a whole. It distracts employees from the company's primary goals, resulting in a loss of both precious time and money. Eventually, customers and stakeholders lose their trust in the company, jeopardizing its success and future growth. Take it from me; it's not worth the risk. Always prioritize the bigger picture and the collective well-being of your customers and stakeholders to ensure long-term success.

Lessons Learned:

The lessons we've learned show that it's vital to prioritize our organizations' shared goals above individual self-interest. It's easy to fall into the trap of prioritizing our areas of influence over the bigger picture, but doing so is a recipe for disaster. As seasoned professionals in the trenches, we know this all too well.

Those in positions of power must recognize that they are responsible for ensuring the organization's overall well-being. Maintaining a big-picture perspective is critical, even as we strive to achieve our goals. Striking a balance between the two ensures we constantly contribute to the organization's greater good.

It's imperative to keep reminding people of the importance of a big-picture perspective. As experienced professionals, we've worked hard to understand the value of prioritizing the organization's common goals above our self-interest. We must continue championing this mindset, even as others may lose sight of the bigger picture.

The road to CRM success is paved with potential pitfalls. Maintaining a cautionary approach is crucial in navigating those challenges with the benefit of our experience. And make no mistake, due diligence is critical in all aspects of our work. We must ensure that individuals in positions of power understand their responsibility in safeguarding the organization's overall well-being.

So, let us embrace this cautionary tone and champion the importance of a big-picture perspective. With our years of experience in CRM, we're well-equipped to navigate the potential challenges that come our way. Let's continue to share our wisdom and help our colleagues approach CRM with diligence, respect, and a focus on the bigger picture.

What Should Have Been Done Differently?

To prevent such adverse effects, business leaders and managers must focus on the company's goals rather than just their self-interests. Transparent communication that sets clear and universal Key Performance Indicators (KPIs) across the company is essential. Additionally, managers should cultivate a culture that values shared values and encourages constructive feedback, promoting collaboration rather than self-serving behaviors.

It is crucial to remember that not everyone is inherently evil. However, it is easy for individuals to get caught up in self-serving events and make decisions that primarily benefit themselves. This scenario highlights the common occurrence of people prioritizing their narrow interests instead of the collective well-being of the organization. To foster a healthy and successful company, emphasis should be placed on joint goals, finding the right balance between personal and organizational interests, and promoting a culture of transparency and collaboration.

22

Too Many Vendors

Justifying The Identification Process involves conducting thorough research, soliciting feedback from relevant stakeholders, and potentially even engaging in pilot testing to determine the application's efficacy and compatibility with the chosen CRM. Negotiating with third-party application vendors can be another significant challenge, as businesses must strike a delicate balance between securing favorable terms and pricing while maintaining productive work. Over the last few years, the CRM landscape has evolved dramatically, with more vendors entering the market and offering increasingly sophisticated solutions. These new options have leveled the playing field, making it both a blessing and a curse for businesses seeking the perfect CRM. While robust competition ultimately benefits all companies in the long run, the sheer number of options available can make the CRM selection process daunting and complex for those responsible for choosing the best solution to drive their organization's success.

Amid this ever-expanding sea of CRM options, many individuals find themselves overwhelmed and tempted to dive headfirst into the world of free trials and vendor presentations, partly due to the barrage of advertising campaigns from leading CRM vendors.

While this approach may seem logical, it can often lead to a lengthy and convoluted evaluation period, with hidden costs and potential pitfalls. One of the primary challenges that businesses face during the CRM selection process is legitimizing the genuine need for a new system. This involves carefully assessing the organization's current pain points, requirements, and long-term goals and determining whether a new CRM is the best solution. This process can be further complicated by the need to evaluate, test, and negotiate with various third-party applications that can enhance the functionality of the chosen CRM.

When evaluating and testing third-party applications for your CRM, you must approach the process with a critical eye and a healthy skepticism. You see, not all add-ons are created equal, and it's crucial to carefully consider whether each application truly adds value to your CRM and aligns with your organization's specific needs.

I've been down this road before, and it can be a bit of a minefield if you're not careful. The marketplace is flooded with third-party applications claiming to revolutionize your CRM experience. But don't be fooled by fancy promises and flashy marketing. Take the time to go deep into the details and needs and truly understand what each application brings.

Ask yourself, Does the application **solve a specific pain poin**t or address your particular challenge, or is it simply really cool? Will it **help streamline** your processes, enhance productivity, or improve the overall efficiency of your CRM? Can this be **done for less time and money** using another method? Would an alternative method be a **bandage or a comprehensive solution**?

THESE ARE the critical questions you must ask yourself and your organization. Remember, you're not just investing money in purchasing the application but also investing time and effort in integrating and maintaining it within your CRM ecosystem.

Beware Of The Vendors

I HIGHLY RECOMMEND CONDUCTING OVERLY-THOROUGH evaluations and testing – don't rely on glossy brochures or online reviews.

DON'T HESITATE **to be "hard and short" with vendors - drive these meetings, and don't let them say whatever they want - this is particularly true for the CRM world in the age of artificial intelligence.**

THERE ARE SO many tools and widgets to solve problems. Take the application for a test drive, put it through its paces, and see how it performs in real-world scenarios. Involve key stakeholders and end-users in the evaluation process, as their insights and feedback are invaluable. Sometimes, despite all your research and careful consideration, you might find that an application isn't the right fit. That's okay! It's better to discover this early rather than invest time and resources into something that hinders your CRM instead of enhancing it.

APPROACH the process with that discerning eye and a healthy dose of skepticism. Consider whether an actual application adds value to your CRM and aligns with your organization's needs. And remember, thorough evaluation and testing are your best friends. Don't rush into any decisions – take your time, involve key stakeholders, and trust your gut. Doing so ensures your CRM has the tools to supercharge your business's success.

IDEA TO CONSIDER - LET me tell you a story about a fantastic leader I once worked with. She had a rather unconventional yet effective approach at the beginning of vendor presentations. She would ask the presenters to physically remove the first five pages (or more) from their paper deck. Yes, you heard that right - she would actually rip those pages out. And you know what? It made quite an impression.

NOW, why did she do this (see *Idea To Consider* above)? It all boiled down to the content on those pages. If they were filled with self-promotional fluff, she wanted to see them gone. She had no time for chest-pounding and vague claims. All she wanted were the facts, the features, and a demo. And let me tell you, **this approach worked like a charm**. It unsettled the presenters, and I mean that in the best way possible. It reminded them that "customers are always right" and that they needed to prove their worth through substance, not empty words. By ripping out those superfluous pages, she sent a crystal-clear message: we're here for a purpose, which is to see the real value your company can offer.

We live in a world of vendors eager to sell us their CRM solutions. They will go to great lengths to dress up their offerings with extravagant claims and flashy marketing materials. But don't be fooled by the glitz and glamour.

When evaluating a CRM vendor or solution, it's crucial to cut through the self-promotional fluff and focus on what truly matters: the features, the functionality, and how it aligns with your business needs. You want to demonstrate how the CRM will help drive your organization forward, not just empty promises.

So, take a page from my former colleague's book and be unapologetically ruthless (but not unkind or rude) regarding vendor presentations. Ask them to skip the pleasantries and get straight to the heart of the matter. Challenge them to demonstrate the key features and functionality that your team needs. Don't be swayed by grandiose marketing claims or glossy brochures - make your decisions based on solid evidence and real value.

Remember, as an executive, your time is valuable, and so is the success of your CRM implementation. Being vigilant and skeptical when evaluating vendors will save you from potential headaches. It will ensure you select a CRM solution that meets your organization's needs and delivers the expected results.

So, the next time you find yourself in a vendor presentation, don't hesitate to channel the spirit of my fantastic leader. Rip out the fluff, demand just the facts, and watch as the actual value of a CRM solution emerges from the clutter

"JUST THE FACTS, mam, just the facts."

23

Case Study: Vendor Situation - Runaway Now

HAVE YOU BEEN A PART OF THIS MEETING WITH A VENDOR?

Scene:

As a seasoned professional, you've likely encountered a vendor who has approached you with an innovative, niche solution that can improve your business processes. After a demonstration, they offer to customize the tool to suit your needs and request that you become a reference for their future sales.

Incident:

You are in this exact situation. A vendor approaches you and shows you an add-on widget with a function that solves your existing problems at a lower price than other options. The vendor is small, innovative, and ahead of the curve with this technology. They need some "wins" to help sell more products, so they offer to customize the tool to suit your needs perfectly and use you as a reference for future sales. It seems like a win-win.

Result:

Unfortunately, doing business with a vendor like this can result in adverse outcomes. While the initial offering may solve a problem at a lower cost, vendors like these are often cash-strapped and may cut corners during development and go out of business shortly. Ultimately, you may end up with a product that needs to be fixed, requires significant adjustments, requires no longer available help, or is downright unusable after initial adoption.

Lessons Learned:

- **Exercise Caution:** Be wary of vendors that promise more than they can deliver or offer discounts that seem too good to be true.
- **Rely on Proven Solutions:** Rather than adopting a niche, innovative solution, rely on proven products with successful track records in the market.
- **Beware of Overpromising Vendors:** Vendors who rely on that all-important next sale may overpromise in the hopes of securing your business, leaving you with an unsuitable product.

What Should Have Been Done Differently?

To avoid unfavorable outcomes, it would be best to:

- **Conduct Extensive Research:** Conduct extensive research before doing business with any vendor to know their track record and performance.
- **Explore Proven Solutions:** Consider proven products that have established user credibility.

- **Understand Their Business Model:** Understand how the vendor's business model works, and if it places too much emphasis on delivering the next sale, be wary.

TRUST ME, RUN AWAY!

As tempting as the offer to adopt an innovative, unproven solution might seem, too many risks are involved. Rather than taking a chance on vendors that may overpromise and underdeliver, it's much safer to research established solutions that have already proved their value. By exercising caution, exploring proven solutions, and understanding vendors' business operations, you can make an informed decision that secures your business outcomes.

This involves conducting thorough research, soliciting feedback from relevant stakeholders, and potentially even engaging in pilot testing to determine the application's efficacy and compatibility with the chosen CRM. Negotiating with third-party application vendors can be another significant challenge, as businesses must strike a delicate balance between securing favorable terms and pricing while maintaining a solid working relationship with the vendor. This requires a combination of diplomacy, assertiveness, strategic thinking, and willingness to walk away from a deal if it does not meet the organization's needs or budgetary constraints.

Once a CRM and any necessary third-party applications have been selected, the implementation process presents unique challenges. Ensuring that the new CRM integrates seamlessly with the organization's existing infrastructure is critical, as is providing comprehensive training and support to help users adapt to the new system. This may involve working closely with the CRM vendor and third-party application providers to address technical issues or customization requirements and developing and delivering targeted training programs for end users. Throughout the CRM selection process, it is crucial to remain vigilant against vendors who may try to push their agenda and persuade businesses to adopt an off-the-shelf solution that does not truly meet their needs. The new CRM must serve the organization, not the other way around, and it is essential to maintain a clear vision of the company's requirements and objectives when evaluating potential solutions. However, the evaluation period can become significantly more protracted when the CRM selection process begins with vendor presentations. A drawn-out selection timeline carries many hidden costs regarding financial investment and the potential opportunity cost of delaying the implementation of a new CRM. Moreover, selecting the wrong product can result in excessive multi-year costs with minimal return on investment, further underscoring the importance of a thorough and well-executed selection process.

The CRM selection process has become increasingly complex and challenging in recent years due to the proliferation of vendors and third-party applications. To navigate this intricate landscape, businesses must remain steadfast in their commitment to identifying and securing a CRM solution that meets their needs while carefully evaluating, testing, negotiating, and implementing any necessary third-party applications. By approaching the CRM selection process with a strategic mindset, a discerning eye, and a focus on the organization's long-term success, businesses can ultimately reap the benefits of a CRM that drives efficiency, collaboration, and growth.

24

Internal Customer Requests For New Technology Tools

While requests for new tools and customizations can sometimes lead to valuable improvements, it's crucial to approach each request carefully. I want to make sure that these requests are not just fancy add-ons that don't align with the organization's overall objectives or fail to offer a genuine return on investment.

Here's a little secret I've learned along the way – sometimes, the functionality that's being requested may already exist within the CRM. It's like searching for your car keys when they're right in front of you! Before implementing a new tool or customization, taking a step back and carefully assessing whether the desired functionality is already available is essential. You don't want to reinvent the wheel when you don't have to.

Another scenario to remember is when a request for customizations is driven by a user's reluctance to adopt existing tools and processes. Change can be tricky, and sometimes people resist it. But your role should never be to say yes to every request but to be that guiding light that helps internal customers understand the potential drawbacks of implementing unnecessary changes.

Remember, it's all about delicately "pushing back" and having those open and honest conversations. Engage in a dialogue with the requesters, understand their underlying motivations, and help them see the bigger picture. Share insights about how implementing unnecessary changes can lead to complications, increase complexity, and hinder the overall user experience.

Your goal is to ensure that the CRM is optimized to meet the organization's objectives without unnecessary clutter or confusion. By carefully scrutinizing each request, offering alternative solutions when applicable, and helping internal customers understand the potential drawbacks, we can strike that perfect balance between progress and stability.

When those requests for new tools and customizations come pouring in, pause, carefully scrutinize their validity, and ensure they align with the organization's overall objectives. And don't be afraid to delicately "push back" and help internal customers see the potential pitfalls of unnecessary changes.

To effectively evaluate internal customer requests for new third-party tools or customizations, organizations should consider the following factors:

Alignment with business objectives; Before entertaining any request, it is essential to determine whether it aligns with the organization's overall goals and strategies. If the proposed tool or customization does not support these objectives, it may not be worth pursuing.

Existing functionality; Thoroughly examine the CRM to determine if the desired functionality already exists. If it does, training users on the existing features may be more efficient and cost-effective than implementing a new tool or customization.

User Adoption; Investigate whether the request is driven by a user's reluctance to adopt existing tools and processes. If this is the case, it may be more beneficial to address the underlying resistance and encourage the adoption of the current system rather than make changes.

Cost-benefit analysis; Conduct a cost-benefit analysis to assess the potential return on investment of the proposed tool or customization. Consider both the direct costs, such as licensing fees and implementation expenses, and indirect costs, such as training and support.

Impact on system complexity; Evaluate the potential impact of the proposed tool or customization on the overall complexity of the CRM. If the change significantly increases complexity, it may not be worth pursuing, especially if it would only benefit a limited number of users.

IF, after evaluating the request based on these factors, it is determined that the proposed tool or customization is not warranted, it is essential to delicately "push back" and help internal customers understand the rationale behind the decision. This can be achieved through the following methods:

Transparent Communication; Communicate the reasons for rejecting the request, emphasizing the importance of aligning with business objectives, minimizing system complexity, and maximizing return on investment. Be openly available to discuss alternative solutions that meet the organization's needs better.

Highlight Cost Implications; Emphasize the financial implications of implementing the proposed tool or customization, including direct and indirect costs. Help internal customers understand that resources are finite, and allocating them to initiatives that offer the greatest value is crucial.

Discuss Time Requirements; Explain the time required to implement the proposed tool or customization, including any necessary training and support. Highlight the potential opportunity cost of diverting resources from other critical projects and initiatives.

Address Limited Usage And Users; If the proposed tool or customization would only benefit a few users, emphasize the importance of focusing on solutions that offer broader value to the organization. Encourage internal customers to consider whether their needs can be met through existing tools and processes.

Offer Alternative Solutions; Suggest solutions to address the internal customer's needs more effectively and efficiently, if possible. This may involve providing additional training on existing CRM features, exploring other third-party tools, or identifying process improvements that do not require system changes.

IT IS essential for organizations to carefully scrutinize internal customer requests for new third-party tools or customizations to ensure that they align with overall business objectives and offer genuine value. The effort will be unbiased and add value by evaluating proposals based on existing functionality, user adoption, cost-benefit analysis, and system complexity. If it is determined that a request is not warranted, delicately "pushing back" through transparent communication, highlighting cost and time implications, addressing limited usage and users, and offering alternative solutions can help internal customers understand the rationale behind the decision and maintain a focus on maximizing the value of the CRM.

25

Conquering The "Innovation Tax" Perception Of CRM

CRM systems have recently emerged as an essential tool for businesses to effectively manage their interactions with customers and prospects. However, some organizations still view CRM as an "innovation tax"—an unnecessary expense that drains resources without providing significant benefits. This perspective can hinder the adoption and effective use of CRMs, ultimately limiting a company's ability to grow and compete in today's fast-paced, customer-centric business environment.

TO FULLY REALIZE the benefits of CRMs, businesses need to conquer this "innovation tax" mindset and make people understand that CRMs are not a tax but a required operational cost. By doing so, organizations can foster a more positive attitude towards CRMs, encouraging employees to embrace the technology and maximize its potential for improving sales, marketing, and customer service efforts.

The Value of Conquering the "Innovation Tax" Perception of CRMs

ENHANCED EFFICIENCY AND PRODUCTIVITY; CRMs are designed to streamline and automate CRM, from tracking leads and managing contacts to monitoring customer interactions and analyzing sales data. By conquering the "innovation tax" perception, businesses can encourage employees to utilize CRMs more effectively, increasing efficiency and productivity.

Improved Customer Satisfaction And Loyalty: CRMs enable businesses to understand better their customers' needs, preferences, and behaviors, allowing them to customize their products, services, and communications accordingly. By emphasizing the importance of CRMs as a required operational cost, businesses can ensure that they are investing in the tools and resources needed to deliver exceptional customer experiences, ultimately fostering greater satisfaction and loyalty.

Increased Sales And Revenue; CRMs can help businesses identify and capitalize on sales opportunities by providing insights into customer buying patterns, tracking the effectiveness of marketing campaigns, and facilitating more targeted and personalized sales efforts. By shifting the perception of CRMs from an "innovation tax" to a necessary investment, businesses can better position themselves to leverage these capabilities and drive increased sales and revenue.

More Informed Decision-Making; CRMs provide businesses with a wealth of data and analytics, offering valuable insights into customer behavior, market trends, and the effectiveness of various sales and marketing strategies. By recognizing the importance of CRMs as a required operational cost, businesses can ensure they are making data-driven decisions that support their growth and success.

Greater Competitive Advantage: A robust CRM can be a significant differentiator in today's increasingly competitive business landscape. By conquering the "innovation tax" perception, businesses can encourage adopting and effectively using CRMs, positioning themselves for greater success in the marketplace.

PROBLEMS ENCOUNTERED When CRMs Are Viewed Negatively Due to Cost

- **Lack of adoption:** Employees who view the CRM as an unnecessary expense may be reluctant to use it, leading to lower adoption rates and inefficient tool use.
- **Reduced ROI:** When the organization does not fully embrace a CRM, the return on investment (ROI) is diminished, as the system's potential benefits are not fully realized.
- **Inefficient processes:** Without a CRM in place or with poor usage, businesses may continue relying on outdated or inefficient processes for managing customer relationships, wasting time and resources.
- **Decreased customer satisfaction:** A CRM helps businesses better understand and serve customers. If the system is viewed negatively, it may not be utilized to its full potential, reducing customer satisfaction and loyalty.
- **Limited collaboration:** CRMs often facilitate better department communication and collaboration. If the system is not embraced, these benefits may not be realized, leading to siloed teams and disjointed efforts.
- **Missed opportunities:** A well-implemented CRM can help identify sales opportunities, improve marketing strategies, and enhance customer service. By viewing the system as a tax, businesses may miss out on these valuable opportunities for growth and improvement.
- **Resistance to change:** If employees view the CRM as a burden or unnecessary expense, they may resist adopting the new technology and processes it brings,

making it difficult for the organization to evolve and stay competitive.

- **Underutilization of features:** When the CRM is viewed negatively due to its cost, employees may not fully explore and utilize its features, limiting the benefits it can provide to the organization.
- **Difficulty in measuring success:** If the CRM is not fully embraced, measuring its success and demonstrating its value to stakeholders can be challenging, potentially leading to further resistance and skepticism.
- **Increased employee turnover:** Employees frustrated with the lack of support for CRMs or who feel their concerns about the system's cost are not being addressed may be more likely to leave the organization, leading to increased turnover and associated costs.

BY CONQUERING the concept of a CRM being thought of as an "innovation tax" and emphasizing its importance as a required operational cost, businesses can overcome these challenges and fully capitalize on the benefits that a CRM can provide. This shift in mindset is crucial for companies to remain competitive in today's fast-paced and customer-centric business environment and to ensure they are investing in the tools and resources needed to drive growth and success.

Paying and Determining Value (Budgeting and Estimating)

CRMS CAN BE VERY expensive and complex, requiring significant time and resources to ensure successful implementation. It's essential to approach CRM projects with careful planning and a clear understanding of the time and expense commitments involved. Lack of proper planning and misunderstanding of these commitments can quickly transform a challenging project into an impossible one.

To navigate the complexities of CRM implementation, it is crucial to provide guidance and establish boundaries around what type of work is realistic and necessary. Rather than simply fulfilling every solution request, assessing the company's needs and striving for alignment with best practices or improvements upon existing processes is essential. This strategic approach helps prevent scope creep and ensures that the final CRM solution meets the company's unique requirements.

The key to successful CRM implementation is finding the right balance between customization and leveraging OOTB features. While customization may be necessary to align the CRM with specific business processes, excessive customization can lead to increased complexity, cost, and maintenance overhead. Therefore, it's vital to carefully evaluate each customization request and weigh the benefits against the potential drawbacks.

Additionally, involving stakeholders from different departments and levels of the organization is crucial for successful CRM implementation. By including representatives from sales, marketing, customer support, and IT, you can foster collaboration, gather diverse perspectives, and ensure the CRM meets the needs of all teams. This collaborative approach helps build buy-in from stakeholders and increases the chances of successful adoption and utilization of the CRM.

Moreover, practical training and change management initiatives are essential to transition to the new CRM smoothly. Many CRM projects fail not because of technical issues but due to user resistance. By providing comprehensive training, addressing concerns, and showcasing the benefits of the CRM, you can drive user adoption and maximize the return on investment.

While CRMs can be expensive and complex, proper planning, clear guidance, and realistic boundaries can help navigate these challenges. By aligning the CRM with best practices, balancing customization needs, involving key stakeholders, and focusing on effective change management, you can increase the chances of a successful CRM implementation that meets the needs of the company and drives improved business outcomes. Remember, the journey toward CRM success requires patience, strategic thinking, and a focus on long-term benefits.

26

Case Study: Allocations Are Needed

Scene:

Setting the Stage for CRM Adoption - Modern CRMs are expensive systems that promise potentially buggy results. The problem is that you don't get a chance to drive the system before you take it off the dealer's lot. Due to this fact, corporations need to agree on a cost with the internal constituents who will ultimately foot the bill. To offset this cost, people want to see what they're getting for their money and how it will eventually pay for itself. This case study will tell the story of a company's attempt to allocate the cost of the CRM not based on license count but on perceived value. I will also discuss a direct charge license model but present an alternative approach with its associated pros and cons.

Incident:

In an era where systems are becoming increasingly costly (and confusing), corporations face the challenge of deriving real value from their investments and proving their worth to people who may never catch up to the curve. Simply acquiring a CRM is not enough; corporations must find ways to make employees want to use the system. If the CRM solution doesn't align with their existing workflows and isn't easily understood, the company will have a severe problem enforcing usage beyond licensing.

Result:

Recognizing the importance of user adoption, the company in our case study aimed to incentivize employees and make them actively engage with the CRM. Instead of solely basing the cost of the system on the number of licenses, they decided to introduce a cost model based on perceived value.

By doing so, they aimed to demonstrate to stakeholders the tangible benefits and ROI of the CRM, thus ensuring higher employee commitment and enthusiasm.

Lessons Learned:

Focus On User Buy-In; Encouraging employees to embrace a CRM requires a well-thought-out plan beyond the initial investment. Consider the system's usability, user training, and ongoing support to ensure that employees use it and find value in it.

Perceived Value-Based Cost allocation; By moving away from a license-count-based cost model and introducing a perceived value-based approach, organizations can demonstrate the direct correlation between CRM and its impact on productivity, efficiency, and revenue generation. This approach helps garner support and investment from stakeholders who may have been skeptical about the CRM's value.

Consider Alternative Licensing Models; While a direct

charge license model is commonly used, exploring alternative licensing models that align with the organization's goals and resources is essential. Understanding the pros and cons of different licensing models will enable companies to select the one that best suits their needs.

Prioritize User Experience; To drive user adoption, the CRM must fit seamlessly into employees' existing workflows. Ensure that the chosen CRM solution is intuitive, easy to navigate, and tailored to meet the organization's and its employees' specific needs.

Modern CRMs may be costly, but their value lies in their effectiveness. The success of CRM adoption lies not just in licensing and implementation but in fostering a culture where employees genuinely understand and appreciate the system's benefits. By prioritizing user buy-in, demonstrating value, and selecting a suitable licensing model, corporations can overcome the challenge of employee engagement and maximize the return on their CRM investment.

27

Case Study: Determining The Value Of Logins

FOOLS MEASURE LOGINS AS A SUCCESS FACTOR FOR VALUE

Scene:

In the world of modern CRMs, determining value has become more complex. The traditional measuring value by analyzing usage and logins has become stale. I used to think that if someone logged into a system frequently, it meant they were getting the most out of it.

But the truth is, just because someone logs in doesn't necessarily mean they fully utilize the system's capabilities. Conversely, someone who seldom logs in may still derive significant value from the system without us even realizing it.

Incident:

Imagine a corporation that heavily relied on the number of logins to measure value. They believed that if employees were frequently logging into the CRM, it indicated high engagement and productive usage. However, the flaw in this approach became apparent when they discovered that some employees, despite logging in regularly, were not leveraging the CRM to its full potential. On the other hand, some employees rarely logged in but were still delivering substantial value through their work, unbeknownst to the company.

Result:

This is the fairy tale result - Realizing the limitations of the traditional 'usage-as-defined-by-logins' concept, the corporation sought new ways to measure the true value their customers received from the CRM. They recognized that the volume of logins alone did not accurately portray the level of engagement, productivity, or effectiveness of the system. As a result, they began to explore alternative methods of assessing value, focusing on the outcomes and impact generated by CRM usage rather than just the quantity of logins.

Lessons Learned:

Focus on Business Processes; Rather than relying on login metrics, focus on how CRMs support the business processes that provide value.

Measuring the Right Metrics; When measuring the value of a CRM, it's essential to consider all metrics that reveal the ROI of the system, not just login metrics.

Engage All System Users; While not all users may log in to the CRM, engaging everyone who will receive value from the system is crucial to understand how it helps them. This is very tricky as many have no idea they are getting crucial information from the CRM or any other "back office" system.

What Should Have Been Done Differently?

To accurately measure the value of a CRM, decision-makers should focus on the following:

Establishing Workflow And Processes; Ensure the CRM supports the business processes that provide value and customize the system to streamline these processes as much as possible.

Choosing The Right Metrics; Instead of measuring logins, focus on other key metrics such as process improvements, data analysis costs reduction, and improved market routes.

Engaging System Users; To understand the true ROI of the CRM, it's essential to engage all users and stakeholders to understand how the system benefits them.

I like to use the analogy that everyone benefits from the expense reporting process, but only some staff enter their own time and expenses. Higher-level staff and executives rarely enter their own time and expenses. If we use the simplified logic of relying too heavily on using login counts, no one could prove the worth of an expense reporting system, which might raise some eyebrows.

Measuring actual logins is not an accurate way to determine the value of a CRM. Instead, focus on how the system supports business processes, measures the right metrics, and engages all system users to understand how it provides value. By doing so, businesses can accurately measure the ROI of their CRMs and make data-driven decisions.

A DIFFERENT SLICE of Reality (really) - Ah, let me regale you with a cautionary tale of obtuse leaders and the perils of clinging to outdated metrics.

Picture this real-world series of recurring events; when it came time to renew the contract every three years, these leaders demanded years' worth of login reports and other metrics straight out of the 1990s. And let me tell you; it was an absolute disaster.

To put it bluntly, this insistence on outdated metrics created an *amazing_flaming_shitshow*. 🔥 **The focus became on substantiating existence rather than considering what could be achieved.** Instead of diving deep into reports that could uncover the true value of the CRM and give the organization a competitive advantage, they fixated on the superficial. They wanted to see "fingers on keyboards" and simplistic, low-value, outdated data—information that had become as old-world and dusty as the floppy disks in the back of your closet.

The result? Nothing of substance was ever proved. Despite years' worth of login reports, it was all just a futile exercise. The organization missed out on the opportunity to explore the potential of what the CRM could truly deliver.

Now, I understand the temptation to lean on familiar metrics, especially when reporting and justifying a CRM contract's renewal. After all, it's human nature to take comfort in what we know. But let me share a hard-earned lesson: relying solely on archaic measures of activity is a surefire way to hinder progress and stifle innovation.

As an executive, your priority should be to extract the utmost value from your CRM investment and gain a competitive advantage in the marketplace. You must shift from simply tracking login numbers and basic activity to measuring outcomes and impact. In other words, it's time to move beyond the "fingers on keyboards" mindset.

Instead of fixating on login reports, consider deeper-diving metrics aligned with your organization's goals and objectives. Explore CRM's value to your sales pipeline, customer satisfaction, revenue growth, and other key performance indicators. Look for ways to measure the return on investment and how the CRM is helping your organization succeed.

To truly unleash the power of your CRM, you need to embrace a forward-thinking approach. Seek reports that analyze customer behavior, identify trends, and uncover insights to drive strategic decision-making. Use analytics to understand your customers better and tailor your marketing and sales efforts accordingly. Get your leaders on board with this mindset to support and champion the transition to a more meaningful evaluation of CRM success.

Remember, the purpose of a CRM is not simply to gather data for the sake of data. It enables your organization to thrive by building meaningful customer relationships, driving revenue growth, and gaining a competitive edge. By shifting your focus from outdated metrics to those that truly measure value, you can ensure your CRM implementation is a resounding success.

So, the next time you face the renewal of a CRM contract, take this cautionary tale to heart. Don't let the relics of the past dictate your future. Embrace the potential of modern metrics and deep-diving reports to unlock the value of your CRM investment truly.

NOTE: Have you found yourself in the unfortunate position of **defending the very existence of CRM** (or CRM support team) within your organization? - It's a frustrating predicament indeed, and one that can suck the life out of even the most enthusiastic CRM advocate. But fear not, for I have a few words of wisdom to share based on my experiences in the CRM world.

WHEN YOU FIND yourself constantly in the trenches, fighting to justify the CRM's existence instead of exploring where it can take your company next, it's time to take a step back and reassess your situation. Here are a few paths you can consider:

Seek A Change From Within; Before rushing to make a drastic decision, take a moment to evaluate whether there are any avenues left to advocate for the CRM's potential. Start by engaging with your organization's key stakeholders, decision-makers, and influential leaders. Share success stories, case studies, and real-world examples of how CRM has transformed other businesses. Illustrate the long-term benefits and competitive advantage that can be gained by embracing CRM to its fullest potential. Sometimes, all it takes is a few open minds to shift the tide within your organization.

Leave Immediately; If your efforts to change the perception of CRM fall on deaf ears and you find yourself banging your head against the proverbial wall, it may be time to cut your losses and move on. Remember, your time and energy are valuable. Suppose you are perpetually stuck in the defensive role, unable to drive the CRM forward. In that case, finding an environment where your expertise and vision are appreciated is best. Life is too short to waste it fighting battles you cannot win.

Wait For New Leaders; Alternatively, you can stay with your current organization and hope for a leadership change. While this may feel like a gamble, it's often surprising how quickly things can shift once new leaders "get it." Changes at the executive level can bring fresh perspectives, a willingness to take risks, and an appreciation for the potential of CRM. However, exercise caution and be mindful of any signs that indicate a lack of commitment or understanding from the new leaders. Waiting indefinitely for change is not a strategy, so set yourself a reasonable timeline for reevaluation.

Ultimately, the decision to leave or stay is a personal one that depends on various factors, including your level of influence, your passion for CRM, and your career aspirations. Just remember, defending the mere existence of CRM instead of exploring its potential can be draining and demoralizing. It's essential to be in an environment that embraces technology and is open to leveraging its power for business growth.

Weigh the pros and cons, and seek confidential advice from trusted mentors and colleagues who understand your challenges. Remember, your expertise and insights are valuable, and there are organizations out there who would eagerly welcome your CRM knowledge and experience.

Take a moment to reflect on your situation and make the best decision with your professional aspirations. Remember, a career in CRM is meant to be inspiring, impactful, and growth-oriented. Don't settle for less, and always strive for an environment that appreciates the value and potential of CRM.

Lessons Learned:

Think beyond surface-level metrics: Logins alone cannot provide a comprehensive value picture. To truly understand the impact of a CRM, we need to look beyond superficial measures and delve into the outcomes and results it generates.

Align Value With Customer Perspective; Value should be

determined from the viewpoint of the end customer, not solely from internal system user activity. Identifying how CRM positively affects customer experience, satisfaction, and business outcomes is crucial.

Measure Tangible Outcomes; Instead of relying solely on logins, focus on the measurable outcomes produced by the CRM. This could include metrics such as increased sales revenue, improved customer retention rates, enhanced productivity, or streamlined processes.

Educate Stakeholders; Help stakeholders understand that simply logging into the CRM does not guarantee value. Demonstrate the need to look at the larger picture and consider the impact and benefits of CRM usage.

We've explored the misconception of value based on the 'usage-as-defined-by-logins' concept and its negative ramifications. However, it's not all doom and gloom. By adopting new ways of measuring value and focusing on tangible outcomes, you can better understand your CRM's impact on your customers and business. It's time to shift our perspective and explore alternative methods to truly assess the value your product delivers beyond mere usage analysis.

28

Case Study: ROI - New Tech

Scene:

Sometimes leaders can be incredibly insightful and make fantastic decisions that lead a company forward in many unexpected ways. However, this is not always the case regarding innovation and adopting new technology. In this scenario, the leaders in question had a narrow perspective and failed to see the potential value of the innovation.

This is the case of approaching leaders that I expected to be insightful; they were only insightful in a ribbon-thin area. They folded in their self-induced floundering when confronted with true innovation and knowledge that could leapfrog them.

Incident:

In this instance, a select group of highly successful leaders in Asia and South America was presented with innovative dashboards that could track business development activity, work-in-progress, and accounts receivable through multiple views and allow email nudges or compliments to be sent. Developed in-house, the dashboards were both free and required no maintenance. However, the leaders' response was less than enthusiastic. One leader asked, "What do I do with that?" while another sarcastically commented that it was "boring." The presenter was then, in essence, dismissed by the leaders with the statement, "Now, what else do you have?"

Result:

The presenter was caught off-guard and left wondering why something that seemed so innovative was so poorly received. As they reflected on the situation, they realized that their highly mechanical processes limited the leaders' perspective. They were so entrenched in their current way of doing things that the new information was jarring and had the effect of proverbially punching them in the face.

So what happened? After many adult beverages, I had to think why something so obvious was so off the rails and completely blindsided the entire team.

Perspective Of The Leader; They were so set in their highly mechanical processes that this new, and f*cking awesome, I might add, information punched them in the face and rocked their world

Caught Flat-Footed; They didn't expect this to be presented and they couldn't figure out how to place this in their existing system. In fact, what was being shown was how to replace the entire system, not a piece.

Suspiciousness of New Things; They had come up through the ranks when none of this was possible. It was like I pushed them off the Titanic and said, "Swim"

Lessons Learned:

Have Patience; It's essential to remember that not everyone will embrace new technology or innovation as quickly as you might hope. Be patient, and remember that change takes time.

Education Is Key; Presenting new ideas or technologies requires effective communication and education so everyone understands the possible benefits. Don't take it for granted that others will be able to connect the dots on their own.

Think Outside The Box; Avoid approaching problems and challenges from a fresh perspective rather than relying on outdated methods.

Understand The Audience: Consider the backgrounds and perspectives of the individuals you are presenting. Make sure that your message is tailored to resonate with them specifically.

Be bold: Don't shy away from pushing the envelope and presenting new and innovative solutions, even if they may be met with initial resistance.

JUST BECAUSE SOMEONE is an insightful leader in one area does not always mean they can capitalize on new ideas or apply those insights to other areas and new technology. As innovative and forward-thinking individuals, it's imperative that we maintain an open mind and are willing to embrace new technology and innovation, even if it means breaking from our current processes.

29

Case Study: Professional Services Firms

Scene:

As CRM professionals, we understand that relying on project timelines provided by service providers, software companies, or project managers can sometimes lead to disappointment and frustration. Although these entities offer a valuable service, it's essential to remember that they have their own objectives and may be incentivized to paint an overly optimistic timeline.

In our work, we've learned from experience that taking timelines at face value can be risky. This is especially true regarding CRM implementation projects, which can be complex and multi-faceted. Considering the stakes involved, it's essential to be diligent and look beyond the timeline to ensure the project's success.

Let's take a moment to think about a realistic scenario. Say you've just received a projected timeline from a service provider. At first glance, it may seem like a great timeline - ambitious but achievable. However, as a seasoned professional, you know that the devil is in the details and that not all factors influencing the timeline are fully accounted for. Software integration issues, unanticipated data migration challenges, and resource availability can all disrupt even the most rigorous schedules.

Incident:

The service provider confidently presents you with a timeline for completing a crucial project milestone. Initially, you may be tempted to trust their timeline, assuming they have considered all relevant factors and have listened carefully to your requirements. However, it is essential to approach timelines with skepticism. Often, those creating the timeline fail to fully understand the complexities of your specific needs, resulting in an inaccurate estimation.

Result:

Trusting a timeline without conducting independent detailed checks can lead to negative consequences. The provider may not have thoroughly understood your questions or the required level of detail. Consequently, their timeline may answer different questions or align with what they assumed you wanted to hear. Additionally, as you delve deeper into the project, you may discover unforeseen technical and business process challenges that cause delays, pushing the timeline further backward.

Lessons Learned:

Independent Sanity Check; Never rely solely on the timeline provided by a service provider, software company, or project manager. Conduct your own independent sanity check, utilizing your expertise and understanding of the project's intricacies.

Ensure Clear Communication; Clearly communicate your requirements and expectations to the provider, ensuring they fully understand your questions and the level of detail needed in their response.

Most importantly; be your own lifeguard and review your desire to hear a timeline that meets your requirements without scrutinizing the details. Ensure it is humanly possible and/or reasonable.

To avoid potential setbacks, it is essential to:

Seek Independent Verification; Take the responsibility to independently verify the accuracy of any timeline provided. Tap into your knowledge and expertise to assess the feasibility and reasonableness of the estimates.

Ensure Understanding; Clearly communicate your requirements to the service provider, software company, or project manager, ensuring they comprehend the questions and the depth of information needed to provide an accurate timeline

Factor In Unknowns; when building the initial timeline, provide some buffer to account for unforeseen challenges and complexities during the project's execution.

MOREOVER, when dealing with service providers and software companies, it's wise to acknowledge that a part of their business model is ensuring they make money - today and in the future. While this does not necessarily imply malicious intent, it's vital to remember that vendors can err on optimism to ensure they win the deal.

Never blindly trust timelines provided by service providers, software companies, or project managers without conducting your independent sanity check. Clarify your requirements, validate timelines through your own expertise, and anticipate potential delays caused by unknown factors. Taking these precautions allows you to navigate projects and avoid unnecessary setbacks confidently.

30

CRM Data Is One Of Your Companies Gold Mines

Company Culture: A Tale of Two Attitudes

Some companies are have a few leaders who have an uncanny habit of behaving like they're the only shining, precious resource in the entire company. Every leadership level (and company) has at least one of these has these information hoarders if not more - who stubbornly believe they're the gold that fills the company's coffers. Their knowledge makes the wheels go around, the ship stays afloat, and the train chug along - Do you get the idea? And the worst part? Frequently, they are unwilling to share the information and data they feel is precious to them and their own proprietary property.

Then there's the other type of company. You know this flavor intimately if you're in one. Here, the staff is the valued resource, the real gold mine of the operation. Every cog and wheel in the machine – the humble worker bees - are considered the true treasure of the organization. This kind of thinking is a godsend in the turbulent seas of corporatism, but alas, it's pretty much as elusive as the fabled unicorn. If you are in one of these, I am envious. Well done to you!

But let's not fool ourselves - and who are we kidding – most companies are a blend of both—sort of a perfectly balanced cocktail of self-centered leadership with just a hint of people-focused policies. And although we're all for balance, it often means you're made to hop from one foot to the other, never quite sure where the weight falls. Does the leader at the very tip-top view their own knowledge as paramount? If so, you will likely be trying to traverse a mountain with a blindfold.

More often than not, many keep their knowledge under lock and key—no sharing and thus limited learning. Slim pickings of the dissemination of wisdom to the diligent foot soldiers who, mind you, get their hands dirty in the trenches. **It is all ironic, isn't it?**

But let's not get all doom and gloom here. It's all part of the corporate environment. The path to success might be steeper, and the climb might suck more than your conventional wisdom would have you believe. Because at the end of the day, it all comes down to you. How you navigate the choppy waters, use your knowledge and skills to make waves (the good kind, mind you), and keep your head above water without losing your mind (and your sense of humor) in the process!

ALTHOUGH DATA IS one of the fundamental drivers of company strategy and forecast, many decision-makers are unaware of how erroneous data collection and poor data upkeep may negatively influence marketing, sales, and the result. Many focus on the future of what they can do without much attention to where they, and their data, have been. When insufficient data creeps into CRMs — whether it's data that's absent, incorrect, placed into the wrong career, non-conforming, or replicated — it will definitely and steadily degrade the system's usefulness to these very executives that are dreaming of the future of this information while driving up the cost of correcting a series of minor information losses.

Marketing With Poor Data

IT'S essential to remember that poor data quality can significantly impact your marketing campaigns. Outdated and inaccurate contact lists can reduce your deliverability rates and increase the likelihood that your emails end up in spam folders. These seemingly minor errors can significantly diminish the impact and effectiveness of your marketing campaigns. They might even distinguish between a campaign generating significant revenue and one barely reaching and engaging your target audience.

To provide an excellent experience to your customers and increase your revenue, it's critical to have high-quality data in your CRM. Access to reliable contact information, such as verified email addresses, is essential in contacting your consumers with news, product specials, and upsell possibilities. Regularly scrubbing and standardizing your data can also help avoid disruptions and lost sales.

I understand that addressing poor data quality is a nagging task, and it can be tempting to push it off when you're busy with other pressing issues. However, I cannot emphasize enough how important it is to make time for it. Poor data quality can lead to missed opportunities, diminished campaign effectiveness, and decreased ROI. Taking the time to enhance and maintain your data quality will undoubtedly pay off in the long run.

Keeping your CRM data clean, accurate, and up-to-date is essential to succeed in your marketing efforts and grow your bottom line. As a mentor and colleague, I advise you to prioritize data quality checks, which will undoubtedly offer significant rewards. Remember, in CRM, even small details like the accuracy of contact information can have far-reaching impacts on your campaigns and revenue generation. Better to invest time and effort now than to suffer the consequences later.

Data quality issues that can lead to CRM failure

A CRM SYSTEM is a powerful tool that can immensely benefit businesses in managing and analyzing customer data, streamlining processes, and improving customer relationships. However, it is crucial to recognize that the success of a CRM heavily relies on the quality of the data it contains. As someone who has witnessed the consequences of poor data quality firsthand, I cannot emphasize enough the importance of avoiding common data quality pitfalls in a CRM.

Incomplete Or Inaccurate Data Entry; One of the most prevalent data quality pitfalls in a CRM is the entry of incomplete or inaccurate data. This can happen for various reasons, from human error to data migration issues. When incomplete or inaccurate data is stored in a CRM, it compromises its effectiveness in producing accurate reports and insights. Moreover, it hinders personalization efforts and can damage customer relationships.

Duplicates And Data Redundancy; Another common pitfall in a CRM is duplicate records and data redundancy. Duplicate records hamper the accuracy and reliability of customer information, leading to inefficiencies and misunderstandings. It also creates confusion among team members, who may waste time and effort contacting the same customer multiple times. Additionally, data redundancy increases storage costs and affects system performance.

Lack Of Data Standardization; Inconsistency in data entry formats, such as variations in capitalization, abbreviations, or naming conventions, is a significant data quality issue. When data is not standardized, performing accurate data analysis and generating meaningful reports becomes challenging. It also affects the ability to segment customers effectively for targeted marketing campaigns. Without proper data standardization, the value and reliability of a CRM can be severely compromised.

Consequences of Neglecting Data Quality Pitfalls

THE CONSEQUENCES of neglecting data quality pitfalls in a CRM can be far-reaching and detrimental to a business. Here are some potential repercussions:

INACCURATE REPORTING; Poor data quality undermines the reliability of reports the CRM generates. Leadership decisions based on incorrect or incomplete data can have disastrous consequences for a business.

Ineffective Marketing Campaigns; Marketing efforts cannot be targeted or personalized effectively without clean and accurate customer data. This results in wasted resources on campaigns that do not resonate with the target audience.

Distrust and Reduced Customer Satisfaction; When customer data is inconsistent or incorrect, it erodes trust both internally among your team members and externally with your customers. Inaccurate information can lead to mistakes, such as sending incorrect offers or addressing customers by the wrong names, ultimately damaging customer satisfaction.

Missed Opportunities; Poor data quality can prevent businesses from identifying valuable insights and opportunities. Lacking a holistic view of customer interactions and preferences, companies may fail to capitalize on cross-selling or upselling opportunities.

MAINTAINING high-quality data in a CRM is essential for its success and the overall success of a business. By avoiding common data quality pitfalls such as incomplete or inaccurate data entry, duplicates and data redundancy, and lack of data standardization, companies can experience accurate reporting, effective marketing campaigns, increased trust, and maximized opportunities. Recognizing and addressing these pitfalls is crucial for the smooth operation of a CRM, nurturing strong customer relationships, and driving business growth. Prioritizing data quality is critical in leveraging a CRM's full potential.

Consequences of Not Avoiding Data Quality Pitfalls

INACCURACY IN REPORTING and decision-making is one of the immediate ramifications of poor data quality. When reports and analytics rely on flawed data, it inevitably leads to ill-informed decisions that harm business strategies and goals.

Furthermore, the implications extend to wasted resources and missed opportunities. Ineffective marketing campaigns and sales efforts resulting from poor data quality not only drain resources but also hinder the potential for revenue growth. The inability to leverage accurate customer data undermines personalized experiences, ultimately reducing customer satisfaction and loyalty. Without a comprehensive understanding of customers, businesses fail to meet their needs and expectations.

Operational efficiency takes a severe hit when data quality is lacking. With poor data and quality come increased manual work, errors, and inefficiencies that plague productivity. This inefficiency negatively impacts operational effectiveness, impeding overall business performance and hindering growth.

However, the consequences do not end there. Neglecting data quality jeopardizes compliance with data protection regulations. Businesses failing to maintain high-quality data expose themselves to non-compliance risk, potentially resulting in significant fines and irreparable damage to their reputation. It requires diligent effort to ensure accuracy, reliability, and compliance. Only by confronting these challenges head-on can businesses safeguard themselves from the far-reaching consequences that poor data quality can inflict.

Addressing Data Quality Pitfalls in a CRM

IN MY YEARS of hands-on experience with CRM systems, I've realized the immense significance of addressing data quality pitfalls to guarantee a successful CRM implementation. For those embarking on this journey, I cannot stress enough the importance of considering the following strategies to ensure the maintenance of high-quality data and fully capitalize on the benefits of CRM.

First and foremost, it is crucial to establish clear and comprehensive data quality standards and guidelines. By setting consistent expectations for data entry and maintenance, businesses can safeguard the accuracy and reliability of their CRM. This ensures that all data entered is consistent and adheres to the established standards.

To enforce data integrity and consistency, robust data validation and standardization processes must be implemented. Utilizing reliable tools and techniques specifically designed for data validation and standardization is paramount. This helps guarantee that only accurate and consistent data is entered and maintained within the CRM.

Regular audits are a necessity in maintaining data quality within the CRM. By conducting regular audits, businesses can identify and rectify data quality issues such as inaccuracies, duplicates, and inconsistencies.

Addressing these issues promptly is essential to prevent the erosion of the system's effectiveness over time. In addition to audits, seamless integration between the CRM and other business software and systems is critical. This integration serves to eliminate data silos and promote data consistency across various applications.

Consequently, businesses can obtain a more holistic and accurate view of customer information, enabling better decision-making and a deeper understanding of their clientele.

Lastly, investing in comprehensive training and education for CRM users is vital. By instilling in users the significance of data quality and providing them with best practices for data entry and maintenance, businesses ensure adherence to data quality standards. Empowering users with the necessary knowledge and skills guarantees consistent data quality management throughout the organization.

By actively tackling common data quality pitfalls through these strategies, businesses can unlock the true potential of their CRM systems. Accurate and reliable data enables sound decision-making, enhancing customer experiences and loyalty. Moreover, a commitment to data quality contributes to the organization's overall operational efficiency. Streamlined processes, improved productivity, and enhanced business performance are just some benefits that will be realized. Diligent investment in data quality is essential. By incorporating these strategies and maintaining high-quality data, businesses can confidently navigate the challenges ahead and set the stage for success in their CRM journey.

31

Data Integration And Clean-Up

The Unapologetically Essential Task: Data Handling

CLEAN DATA IS NOT JUST nice to have. It is an absolute necessity for the success of your CRM efforts. Without clean data, your system will not work correctly, and you'll waste resources on fixing issues instead of utilizing your CRM software to run your business efficiently. So, let's dive into how to handle data effectively.

When discussing data, I refer to the information you gather from your customers, prospects, and various business systems. Data handling is like the beating heart at the core of your CRM effort. It is the foundation that everything else relies on. Your CRM system will falter if your data is not clean, accurate, and reliable.

So, what exactly does clean data mean, and how can you achieve it? Let's explore the key areas you must prioritize:

Data Structure; Properly organizing your data is crucial. Ensure consistency, uniform formatting, and ease of understanding. A well-defined data structure eliminates confusion, reduces errors, and enhances efficiency. Take the time to establish an explicit data structure and enforce its adherence within your organization.

Data Quality; The accuracy and reliability of your data are paramount. Regularly audit your data, address duplicates, and fix inconsistencies or inaccuracies. Poor data quality will hinder decision-making and customer service, undermining the effectiveness of your CRM efforts. Remember, garbage in, garbage out.

Data Security; Protecting customer data and maintaining trust are top priorities. Implement robust security measures, including access controls, encryption protocols, and regular security audits. Data breaches can have disastrous consequences for your business. Treat data security as a non-negotiable aspect of your CRM strategy.

Data Trust and Reliance; Your stakeholders, including customers, must completely trust the data within your CRM system. Keep your data up-to-date, accurate, and relevant. Foster a culture of relying on CRM data for decision-making rather than personal biases. Making data-driven decisions enhances the overall success of your CRM initiative.

ALL PARTICIPANTS, customers, and clients must understand that clean data is not just a fancy or optional feature. It is an imperative quality, a fundamental requirement for your CRM and business's success. Please don't waste time with those who dismiss its importance. Educate and emphasize the significance of clean data. Anyone who fails to grasp this fundamental understanding should be prevented from participating. It may sound harsh, but allowing them to influence decisions can quickly lead you into a mire of ineffective CRM practices.

So, be diligent in handling your data. Treat it as the lifeblood of your CRM efforts, and you will see its positive impact on your business.

NOTE: I suggest you implement a reward vs. consequences for following or breaking best practices.

REMEMBER, you're not doing them a disservice by making a mountain out of this molehill. Making rewards for good practices or process change can be a powerful tool. **Remember you're saving future headaches and heartaches by drilling into your team and clients the magnitude of the importance of Clean Data**. A mess-free CRM operation is the prize waiting for this harrowing journey's end. Equipped with the unshakable understanding that Clean Data is the foundation upon which our CRM success stands – non-negotiable, essential, and unapologetically central to our collective cause.

Close Scrutiny On Clean Data

LET'S not kid ourselves here. All future tools, needs, triggers — you name it — are hinging on clean data. And crap, is it a slippery slope! There are more ways to muck it up than you can believe and just a handful of ways to make it sing.

First order of business in our no-nonsense big data management book: pay close meticulous attention to your data. Handpick what makes the cut and toss aside the rest. It's about as glamorous as watching paint dry, but believe me, it's the vitamin C shot your goals need.

Take a moment to consider the torrent of data running through the veins of your company—system after system acquiring, storing, and reporting on this lifeblood if you're lucky. Many systems execute transactions based on the same data, hoarded like trophies in multiple locations.

This overlapping data, or the monster lurking under your bed, can either be a lever for success in the data integration process or a self-destruct button that detonates your future success (short and long-term, thank you very much). Now, don't pull your hair out just yet. Redundancy can be a good thing (said no one ever - until now). Yes, reliable yet redundant information, as much as it sends us into a tizzy, can be crucial in ensuring that your data is lined up, neat as a pin. It offers that warm, fuzzy feeling of reassurance that the system is assessing the same data set, irrespective of it roaming around in distinctly different systems.

Hang on, however. This process, our so-called redundancy rescue, can morph into a monster and start spewing false confirmations of data confidence. Before you know it, you're knee-deep in a shitload-data-quagmire, with what you thought was redundant data turning out to be similar but not quite the same. Stay tuned for more riveting revelations on this topic in the forthcoming, in no way sleep-inducing chapter, "Case Study: Let's Not Un-Dedupe the Deduped Data." I promise it's full of twists, turns, and tangles — a veritable rollercoaster ride (sorry, I couldn't resist throwing that one in!).

Data Management, Knowledge Management, and Data Hygiene are jargon we often hear when dealing with CRMs. There's nothing quite like advocating the necessity of pristine data; if you are good at it, you might not even sound like too much of an a-hole. But everyone should know you can't have a reliable system without it. So if they think you are an a-hole, I can confirm they are too. The question may quickly become, "Which data is the correct data?" but more on that later.

Sure, we could drop the mic and call it a day there, but where's the fun in that? Instead, let me paint a picture. You have a technical solution that scrubs away dirty data like there's no tomorrow. Now what? How do you have a process that aligns with this data-cleansing marvel? Are you ready? People need to own their data. There's no sugarcoating or dancing around this one; it's the one-way ticket to the CRM Promised Land. But how you ask? Let me explain with a few possible ideas:

Assigned Data Stewards; Rock the boat, appoint your most trustworthy and meticulous Data Stewards—those who are equal parts data wizards and organizational Picassos. Their sole mission? To oversee and take accountability for their designated data segments like the crown jewels. Remember, a firm sense of ownership leads to more careful data handling and reporting.

Data Management Training; Unleash the secret weapon of training and education. Provide your team with the knowledge and skills they need to wield their CRM powers. A data-savvy team makes for a well-oiled CRM machine that can take on any challenge. We're talking comprehensive rundowns on data quality, capture, analysis, and more. The more they know, the fewer loose ends you'll have flying around.

Clear Communication Channels; Keep the chain of command and communication straightforward because information silos are the stuff of nightmares. Foster an open, collaborative atmosphere where anyone can raise red flags about data concerns or seek advice. A receptive and transparent organization paves the way for better data management.

Ongoing Data Audits; Set up a rigorous, unapologetically thorough audit process to give your Data Stewards bated breath and sweaty palms. I am not joking; this is the only way to identify inconsistencies, issues, and gaps in your data. Regular audits will keep your data in check and help assess whether your team is taking ownership of their share.

Feedback and Rewards; Last but not least, create an environment where feedback (both positive and constructive) is appreciated and where hard work and vigilance don't go unnoticed. Reward those who demonstrate impeccable data ownership, and you'll find that these shining examples will encourage and motivate others in their quest for CRM data nirvana.

TOGETHER, your team and divisional allies should be able to master the art of CRM data management.

32

Collection And Integration Of Data

Regarding customer data, four key components form the foundation of a strategic approach: acquisition and integration of data, management of data, analyzing data, and activation of data. These components work together to help organizations harness the power of their customer data and drive meaningful outcomes.

Let's break down each of these components to understand their significance in the broader context of customer data management.

ACQUISITION AND INTEGRATION OF DATA; The first step in any customer data strategy is acquiring and integrating data from various sources. This involves collecting data from different touch-points, such as website interactions, mobile apps, social media, and offline interactions. The goal is to create a holistic view of each customer by consolidating their information into a centralized database.

Management Of Data; Effective data management becomes crucial once the data is acquired and integrated. This involves organizing and maintaining the data in a structured manner, ensuring its accuracy, completeness, and security. A CRM system is pivotal in managing customer data, serving as the engine that drives sales, marketing, and customer service. The CRM database serves as the fuel that keeps everything running smoothly.

Analyzing Data; Customer data provides valuable insights and enables organizations to make informed business decisions. By applying data analytics techniques, businesses can uncover patterns, trends, and correlations within the data, allowing them to better understand customer behavior, preferences, and needs. These insights are crucial for developing personalized marketing campaigns, improving customer experiences, and identifying growth opportunities.

Activation Of Data; Activation refers to using customer data to drive meaningful actions and initiatives. Organizations can leverage the data to recruit new customers, run targeted marketing campaigns, and generate revenue by activating it. Customer data platforms (CDPs), data management platforms (DMPs), and data exchange platforms (DXPs) are widely used software applications that facilitate the intricate process of collecting, consolidating, and activating consumer data.

A CUSTOMER DATA platform (CDP) is a marketer-managed technology that creates a comprehensive, continuous image of customers. It enables marketers to effectively leverage this data to activate campaigns and other marketing initiatives. CDPs typically manage first-party customer data, which includes personally identifiable information (PII).

On the other hand, data management platforms (DMPs) focus on collecting and retrieving campaign and consumer intelligence, primarily for digital marketing campaigns. DMPs work with anonymized data, including third-party information such as cookies, machine IDs, and IP addresses. Given the benefits of working with owned data and the increasing importance of data protection and compliance rules like GDPR, many organizations prioritize CDPs and first-party data.

A well-executed customer data strategy relies on effectively collecting, integrating, managing, analyzing, and activating customer data. By implementing technologies like CDPs and DMPs, organizations can unlock the full potential of their customer data and drive impactful outcomes. Staying informed about evolving data protection regulations and compliance rules is essential while leveraging customer data to achieve your business goals. So, invest in the right tools and strategies to empower your organization in the age of data-driven decision-making.

33

Migrating Data

It is a matter of great intrigue to witness the overwhelming majority of individuals migrating data from one or multiple systems to a new platform without carefully considering the underlying motivations. What the f*ck did you just say Hudlow? How can data have motivation? Easy, everyone is so into their old data that it has become taboo to even mention its worth in a new system. In my view if something is taboo, then it is a prime target for improvement as it is passed over so quickly that people don't even consider it a choice, and we always have a choice. The real question should not be whether to move the old data but rather how much to move. It's a choice that requires careful consideration and understanding of the potential benefits and/or issues of moving old data. For instance, transferring irrelevant data can bloat the new system and make it more challenging to utilize effectively.

It's also essential to evaluate the quality of old data and identify any inaccuracies or redundancies. Cleaning and consolidating old data can optimize new system performance while reducing maintenance costs.

So, what's the solution? The answer lies in careful consideration of the potential benefit of the old data. Determine the data fields that are necessary and relevant to your current and future business objectives. Prioritize the clean-up of high-value data fields, and consider retiring or archiving the rest.

Old data may have a stigma attached to it, but this does not mean it should be ignored or dismissed outright. Take the time to evaluate the quality and relevance of your old data before making any decisions. By carefully selecting and migrating only high-value data from your legacy system, you stand to gain valuable insights that can drive business success in the future.

It is a prevailing assumption among many that all data, spanning the entirety of the organization's existence, holds equal importance and must be fully retrievable in the new system. However, addressing this prevailing perspective and considering the factors surrounding this decision-making process is essential.

When contemplating data migration, performing a thorough analysis considering various critical factors is imperative. The cost implications of migrating all available data should be meticulously examined. The endeavor to convert every piece of data demands significant resources, including financial investment, to ensure its successful implementation. It is crucial to evaluate the economic viability of such an undertaking, considering the potential impact on the project budget and the overall return on investment.

Furthermore, the proficiency and skill set of the team responsible for executing the data migration must be seriously considered. The conversion of vast quantities of data necessitates a specialized skill set and expertise, particularly in handling data transformation processes and ensuring data integrity throughout the transition. It is paramount to assess the skill level and proficiency of the individuals involved and, if necessary, augment the team or seek external support to ensure a smooth and efficient migration process.

Time is yet another critical factor that must not be overlooked. Migrating an extensive volume of data requires a considerable amount of time and effort. The process involves thorough planning, extensive testing, and rigorous quality assurance measures to guarantee the accuracy and integrity of the migrated data. It is imperative to evaluate the time available and allocate resources accordingly to avoid potential delays or disruptions to ongoing operations.

Moreover, it is vital to acknowledge that the decision to migrate all data is not solely a matter for the project team or system administrators. Instead, it affects a broad spectrum of stakeholders, including end-users and customers. It is essential to consider the implications of data migration on their user experience, ensuring that any disruptions are minimized and their needs are effectively addressed. The perspectives and expectations of these stakeholders must be considered to ensure a smooth transition and maintain high levels of satisfaction and engagement.

Before committing to comprehensive data migration, engaging in a thoughtful evaluation of the benefits and drawbacks of such an approach is vital. This requires a thorough assessment of the value and relevance of each piece of data, considering its impact on business operations and strategic decision-making. Taking the time to classify and prioritize data based on its significance will enable a more efficient and targeted migration approach, focusing on the information that truly adds value to the organization.

The decision to migrate data from one or multiple systems to a new platform is a venture that demands careful consideration and evaluation. It is vital to assess the costs, skills, and time required to convert and integrate the data effectively while also considering the perspectives and needs of all stakeholders involved. By embarking on a comprehensive assessment encompassing financial feasibility, skill requirements, time constraints, and stakeholder considerations, organizations can ensure a well-informed decision that minimizes potential risks and maximizes the benefits of data migration.

BUSINESS OBJECTIVES AND NEEDS; Align your decision with your organization's goals and needs. Consider how retaining or discarding existing data will impact your business operations, decision making, and future growth. It's crucial to balance leveraging the insights from historical data and avoiding unnecessary clutter that could hinder progress.

Data Quality; Evaluate the quality of your existing data. Ask yourself: Is the data accurate, complete, and relevant to your current and future needs? If the data is outdated, incomplete, or unreliable, it might hinder decision-making and compromise the effectiveness of your new system. However, if the data is reliable and still holds value, preserving it can offer a foundation for analysis and trend identification.

Regulatory and Legal Obligations; Consider any legal, regulatory, or compliance requirements about your data. Strict guidelines govern data retention in specific industries like finance or healthcare. Not complying with these regulations can result in harsh consequences. Ensure you know any particular retention periods or data disposal protocols that apply to your business.

Cost and Storage Capacity; Keeping data comes with storage and maintenance costs. Consider the financial implications of retaining large volumes of data. Assess your storage capacity and evaluate the scalability of your infrastructure. Balancing valuable data and managing the associated costs effectively is essential.

Data Security and Privacy; Protecting sensitive data needs to be a priority for any organization. Evaluate the security measures in place for storing and accessing your data. If your current data storage methods lack the necessary security protocols, it is advisable to migrate to a new system that offers enhanced data protection and privacy features.

Data Accessibility and Integration; Consider how easily you can access and integrate your existing data into the proposed new system. Evaluate the compatibility and interoperability between your current data format and the new system's requirements. In some cases, data migration can be complex and time-consuming, and the effort required might outweigh the value of retaining the data.

BY CAREFULLY CONSIDERING THESE FACTORS, you can arrive at an informed decision regarding whether to keep or discard existing data. Finding the right balance that aligns with your business objectives, data quality requirements, regulatory obligations, and financial considerations is essential. Consulting with data and technology experts can also provide valuable guidance and insights tailored to your situation. Make your decision based on a **thorough analysis of the pros and cons (see list at the end of this section)**, and always prioritize your organization's long-term success and growth.

It will be a bumpy road filled with frustrating choices and input from multiple people. To navigate this journey successfully, you'll need to address the historical issues that have plagued your data, attempt to rectify those issues to keep them out of the new system, convert processes and reports, and get the approval of the existing data "owners."

IMPORT - DO NOT F*CK **Up - Hidden Tip -** It is imperative to set expectations in this step. When you move old data, the original owners also assume that old processes and reports will still exist and that the gremlins from the past will be completely vanquished. Make what will be and what will not and avoid this problem avoid that at all costs.

SETTING realistic expectations for its actual use is essential when transitioning to a new system. As an experienced mentor bearing the battle scars of making wrong decisions, I can tell you that users often assume that the old system's features will be perfectly replicated in the new one. They may think their current processes will quickly and flawlessly align with the new system's capabilities. But the truth is, merging systems always require retooling and potential data realignment. It's not as easy as copying and pasting the old processes.

All involved need to understand that it's essential to understand that one new system won't blend all pre-existing process paths into one ideally improved process. The overall process should be improved to serve the needs of more people better or provide more comprehensive information. However, it's not uncommon for one person's approach to morph into something equally (if not more) complex. Delicately handling these situations is crucial for a successful project outcome.

Keep in mind that it's essential to communicate honestly and effectively with all stakeholders involved. Providing clear explanations and real-world examples can help set more realistic expectations in the future. Balancing cautionary advice with practical solutions will help keep everyone's best interests in mind and ultimately lead to project success.

SO, how do we approach this migration journey like seasoned pros? Here are a few more tips and insights from my experience in the field:

SET EXPECTATIONS; Make it crystal clear from the start that the migration process will not bring back the ghosts of the past. Communicate that old processes and reports might not have an exact equivalent in the new system. By managing expectations upfront, you'll avoid disappointment and prevent unnecessary work.

Prioritize And Assess; Take a good, hard look at your existing data. Identify the key historical issues that have been causing problems. Determine their impact on your current operations and decide which issues must be addressed urgently and which can be left behind. By prioritizing and assessing, you can focus your efforts where they're most needed.

Rectify And Clean; Before migrating, invest time and effort in rectifying the data issues you've identified. Clean up your data by removing duplicates, correcting inaccuracies, and ensuring consistency. This will help you maintain data integrity and avoid transferring problematic information to the new system.

Convert With Care; When converting processes and reports. Take the time to understand the capabilities and limitations of the new system and adapt your processes accordingly. Remember, change is an opportunity to improve, so embrace it and find ways to streamline and enhance your workflows.

Involve The Right People; Get the approval and support of the existing data "owners." Collaborate with them to understand their needs and concerns. Involving the right people from the start fosters a sense of ownership and ensures a smoother transition. Plus, it's always helpful to have allies on your side when challenges arise.

Test, Test, Test; Don't underestimate the power of testing. Thoroughly test the new system and processes before going live. This will help uncover any issues or gaps that need to be addressed. Conducting rigorous testing minimizes the chances of encountering significant hiccups.

REMEMBER, migrating can be challenging, but you can navigate it successfully with careful planning, attention to detail, and open communication. Setting clear expectations, addressing historical issues, adapting processes, and involving the right stakeholders pave the way for a smoother transition. And most importantly, don't forget to celebrate the small victories along the way—those little wins keep us motivated and help us overcome the bumps in the road. Safe travels on your migration adventure!

Migrating data requires strict rules (see more in the previous chapter, "Being Strict") that must be created, and expectations must be continually monitored and corrected. I have found that it seems human nature for people to expect that a new system will not only have their old information in it but all negative aspects of the data itself' will have been cured. Additionally, the improvements to their related reports and dashboards, regardless of age and future usefulness, will be part of the "grade" those individuals give to your entire migration effort.

When considering implementing a CRM system, assessing the potential pros and cons is essential. Potential advantages include reduced time and effort needed to establish specific data parameters, more immediate access to existing information, and user-friendly functionality to facilitate easier promotion and selling. However, implementing a CRM may also lead to inherent risks, such as existing data issues that will not disappear without cleaning or deduping, which can also pose potential risks to data integrity. There may also be conflicts with existing data formatting requirements, meaning filters, drop-downs, and other settings may have to be duplicated or recreated. This process often necessitates the involvement of third-party experts to ensure that the implementation meets the set standards. Therefore, it's crucial to undertake due diligence, recognize the potential pitfalls, and weigh the benefits and risks before implementing a CRM system.

34

Case Study: Let's Not Un-Dedupe The Deduped Data

Scene:

A bustling company with a global presence, its divisions spread worldwide. Each division, functioning almost like a separate entity, has its own way of capturing and storing contact information. Some rely on databases, others on spreadsheets, and every division has developed its unique system to manage customer data. The result is a scattered landscape of information, like puzzle pieces across the globe.

Imagine this company's ambitious endeavor to consolidate all that contact information. The goal is admirable - to create a unified, comprehensive view of their customers, streamline processes, and enhance decision-making based on accurate and up-to-date information. However, the task is far from simple.

Incident:

In the months leading up to the established go-live date, the project team at a competent and mature company was working tirelessly to gather data records from various sources. Recognizing the importance of a smooth import process, many groups involved in the system's future use diligently created vast spreadsheets of contact records, aligning them to an established format. With more than 100,000 contact records submitted, both the project team and the hired 3rd party installer were expected to deduplicate and import them into one cohesive and organized system - a modern CRM.

The atmosphere was optimistic as everyone believed their data submissions would be transformed into shiny, new, and polished resources for future business development efforts. The team was overjoyed at the prospect of consolidating all their valuable contact information in one place. However, despite providing templates and clear guidelines, many individuals submitted spreadsheets that were almost, but not quite, up to the required standards. The level of quality and completeness varied, leaving room for error.

Result:

A Catastrophic "Shitshow"

The consequences of these compromises on data quality and completeness became apparent too soon. The project team focused on staying on schedule and ensuring customer satisfaction, decided not to burden the practitioners by rejecting submissions that didn't meet the established guidelines. Instead, the spreadsheets were accepted without validation or the opportunity for practitioners to remove personal contacts that didn't belong in the CRM. It seemed easier to forge ahead without insisting on pre-cleaning the contact information or relinquishing perceived control over their own contacts.

As a result, the practitioners mistakenly assumed that their data was being automatically cleaned, normalized, deduplicated, and stripped of personal contacts and other private information. Unfortunately, reality struck when they discovered their assumptions were far from true. Frustrations grew, and the blame was shifted to the implementation team and all those affiliated with the project, accusing them of failing to "fix" the data before migration.

Lessons Learned:

The Importance of a Data Effort

One crucial mistake made by the implementation team was not insisting on a parallel data effort alongside the CRM implementation. By neglecting to prioritize the creation of a quality information database, the team unknowingly set the stage for years of unhappiness and the eventual downfall of using the system as a reliable tool for business development.

The lack of attention to the data aspect of the implementation ensured that every user continued to rely on their original tools, such as Excel, for day-to-day operations. The modern CRM became a one-off solution, primarily used for more targeted work. This reliance on individual tools was driven by the perception that they were more secure and faster.

The Unrepairable Fog

So, what exactly happened to the data loaded into the modern CRM? The non-uniformity of the data, combined with its varying levels of completeness and the presence of discrepancies between similar or identical records, created a dense fog that ultimately rendered the information untrustworthy and unrepairable.

Visualize the deterioration of the data by imagining a document being copied repeatedly using the most recently produced copy as the new master. After a few cycles, you even fax the document to yourself. Now, envision the quality of that document. Would you use it? Certainly not. And just like that, trust in the tool and those who advocated for its use crumbled.

In this case, data was submitted from multiple sources, resulting in duplicative and unique records. Any contact record could be simultaneously correct and incorrect, depending on the submission source. The overlaying of these conflicting records resulted in a hazy and unreliable view of the information essential for carrying out practical business activities.

What Should Have Been Done Differently

It is essential to take the following steps when implementing a CRM:

Prioritize Data Quality; Insist on a dedicated data effort that runs parallel to the CRM implementation. Building a robust and reliable information database should be a core focus from the very beginning.

Establish Clear Guidelines; Communicate the required data quality and completeness standards to all stakeholders. Provide templates and ensure that submissions adhere to these guidelines.

Validate and Verify Data; Allow practitioners to review and validate their data submissions before import. Provide them with the opportunity to remove any personal contacts or information that should not be included in the CRM.

Conduct Rigorous Testing; Before going live, thoroughly test the import process, deduplication mechanisms, and data validation processes. This will help identify and rectify any issues or inconsistencies before they cause significant damage.

Communicate and Educate; Ensure that everyone involved understands the importance of data quality and their responsibilities in maintaining it. Provide training and ongoing support to encourage using the CRM as the primary tool for daily activities.

BY INCORPORATING these measures into the CRM implementation process, you can avoid the pitfalls and catastrophic consequences that befell the unfortunate company in this cautionary tale. Remember, the foundation of any successful CRM lies in its data, so invest the time and effort to build a robust and reliable information database.

Part II: Implementing And Going Through The Setup

In the second section of this book, I dive into crucial topics that often arise when configuring and setting up a CRM for practical use. Whether you're tweaking a brand-new CRM, customizing an existing one, or configuring a bolt-on for your core system, this section explores topics that demand extra attention. It's essential to recognize that neglecting these areas can disrupt your plans and cause unforeseen stress if not adequately addressed.

During this phase, it is essential to consider everything

mentioned, in addition to insights you've gained from your own unique circumstances. The key takeaway is that the topics discussed in this section extend beyond the setup phase alone. They can surface at any point throughout the entire CRM lifecycle, from initial implementation to ongoing maintenance and future upgrades. Equipping yourself with a deep understanding of these topics early on will better prepare you to tackle unexpected challenges that may arise down the line.

Remember that our goal here is to empower you with knowledge and insights to make informed decisions while proactively anticipating potential roadblocks. It is vital to contemplate these topics earnestly, take them to heart, and incorporate them into your CRM strategy. Doing so will help you avoid unnecessary stress and enable you to tackle any hurdles that may come your way.

Experience has taught us that due diligence and meticulous planning are the key ingredients for CRM success. By paying close attention to these critical areas, you'll confidently navigate the CRM landscape, knowing you are well-prepared for any challenges that may arise during your CRM journey.

1

Not "Sunsetting" Redundant Systems

FEAR OF PUTTING YOUR EGGS IN ONE BASKET

Warning: Not sunsetting, or an agreed upon path to sunsetting, a legacy and quasi-duplicative tool with the new CRM is the penultimate kiss of death.

I CANNOT STRESS ENOUGH the importance of sunsetting older technology that is no longer needed - even technology that is still usable, but duplicative, needs to go! The dangers of not doing so are numerous and can have disastrous consequences. Despite the allure of shiny new technology with better capabilities, many organizations fall into the trap of not letting go of their older systems. This can lead to confusion, miscommunication, and operational inefficiencies that can significantly impact the bottom line.

First and foremost, having both older and newer technology operating simultaneously can cause a great deal of confusion for users. It's essential to note that people are creatures of habit and often resist change. Keeping the older system around can encourage people to stick to what they know and not take full advantage of the newer, better technology. This can lead to decreased productivity and user frustration, which can have a ripple effect throughout the organization.

Another danger is that maintaining both systems requires additional resources and expenses. Paying for both the old and new technology can be costly and can hinder an organization's ability to invest in more modern systems that would benefit its operational efficiency. One must also remember that maintaining outdated technology also comes with its challenges, such as security breaches, obsolete hardware, and software maintenance costs, all of which can compound over time.

Additionally, having two systems operating at once can hinder best practice adherence. In today's highly competitive business landscape, staying ahead of the curve requires a strict commitment to best practices and adopting the latest technologies. Keeping an older system around can put an organization at a significant disadvantage as it can impede the integration of the newer system's best practices. This can result in higher costs, slower adoption, inefficiencies, and decreased productivity.

Another significant issue of not sunsetting older technology is becoming stuck in legacy systems. An organization that resists the adoption of newer technology can become trapped in an endless cycle of maintaining outdated systems. The longer this goes on, the harder it becomes to break free, as there is always a fear of disrupting the status quo and the potential for unintended consequences.

The dangers of not sunsetting older technology despite having newer, better technology that can replace it and comes with enhanced capabilities are significant. Operating both systems simultaneously will create confusion, impede best practice adherence, and increase expenses, among other negative consequences. Staying ahead of the curve requires organizations to adopt best practices and stay up-to-date with the latest technology trends. To do so, it's essential to sunset outdated systems and invest in the latest technologies that support organizational goals and objectives. Due diligence is crucial in this regard, and organizations must weigh the risks and costs of maintaining older systems against the benefits of adopting newer technologies.

Let me share a cautionary points about the perils of clinging to outdated systems. As a seasoned professional with years of experience in the trenches, I've witnessed more system dreams go up in flames because of precisely this issue.

As a start there are common excuses VIPs often give to justify keeping their outdated systems alive: (these are excuses, not reasons - in other words, they are **BS** 🐮). In turn, these reasons become edict for their staff (which may be huge) to continue the “old ways”

"It was made years ago, is paid for, and is working”;While it may seem valid, relying on an aging system can lead to missed opportunities and inefficiencies. Just because something works doesn't mean it's optimal for your business's growth and success.

"It is intertwined too deeply to redo": This excuse often stems from a fear of disrupting day-to-day operations. However, modern CRMs are designed to integrate seamlessly with existing systems, mitigating the risks associated with a full-scale migration. - This is also sometimes referred to as the Big Hairy Beast Effort 🦍, which scares everyone away.

"Everyone is used to using it": Familiarity should not be a compelling reason to retain an outdated system. With proper training and change management strategies, your team can adapt to a new CRM with enhanced capabilities and efficiency.

"We are too busy to shift gears": The busyness of day-to-day operations should not be an excuse to neglect the long-term benefits of a modern CRM. Investing time in implementing a new CRM can streamline processes, improve productivity, and alleviate the burden of busyness.

IN ADDITION to these common excuses, secret reasons often lurk behind the resistance to change:

Emotional Attachment; Some individuals, who may have been interns or involved in creating the old system years ago, develop emotional attachments to their creation. It becomes difficult for them to let go and embrace a newer, more powerful CRM.

Sense Of Control; Some see the old system's control as part of their domain, their kingdom. They enjoy the authority and control it grants them and are reluctant to relinquish it.

Reporting Structure; The staff supporting the old system sometimes reports directly to these decision-makers. This arrangement creates a conflict of interest, incentivizing them to preserve the status quo.

Fear Of Exposing Flaws; There might be a hidden fear that modernizing the system will reveal the deficiencies of the old one. Those in control may not want anyone pulling back the covers and exposing the flawed foundations.

Laziness: Sadly, some simply resist change out of pure laziness. The effort required to transition to a modern CRM seems overwhelming, so they ignore the benefits it could bring.

HOWEVER, I must emphasize that very few of these reasons hold, and none of them should hinder your company's determination to embrace an excellent modern CRM. The benefits of a current CRM, with its advanced features, data analytics capabilities, and increased efficiency, far outweigh the resistance to change. It is imperative to grind away these excuses as soon as possible and proceed to implement a single, unified CRM.

By overcoming these obstacles and selecting a modern CRM path, you can harness the full power of a reliable CRM to streamline processes, enhance customer relationships, and drive business growth. Don't let excuses and resistance stand in the way of progress. Invest time and effort into due diligence, evaluate the modern CRM options available, and embark on a journey that will propel your business forward. Trust me, avoiding these pitfalls will save you from the regret and headaches of outdated systems.

This point might be the most important in the book, do not

proceed with any effort to implement a new system if you also don't have an agreed upon and approved plan to remove/uninstall/stop the use of any process and associated methods it will be replacing. This means that everyone agrees, from every department and geography, and most important, the assholes that are waiting to throw a wrench in the mix at the 11th hour.

My points in the case study of "Dueling Systems" is frequently lost by the new (and usually better) system(s). Even worse, many existing smaller, less widely used methods and processes are being used than you realize. Each and every old process and system must be phased out if your effort is to be a success. It is that simple. At the core of the issue is that the old/legacy/existing system(s) have become routine/2nd nature to use, and it makes them difficult to remove and replace for some; additionally (and even worse) is that the creators of that existing system still may work at your company and they have an emotional attachment to it. All of these people need to be identified immediately. They must be listened to and convinced of the new path forward with care.

This concept often gets lost in the shuffle: without a plan to retire your old system, your new system is doomed. **Yes, it's a fact.** And just because it's frequently overlooked doesn't make it any less accurate. If you're introducing a new CRM system and not firmly planning to discontinue the old system(s), this might be the death shot for your new one. <u>It's not just possible; it's incredibly likely.</u>

REPEATING A MESSAGE - WITHOUT AN ACTIVE, agreed upon, and aggressive sunset plan, your new system is a goner. **Fact.**

THIS CONCEPT ACKNOWLEDGES that change is a constant and unavoidable part of life, both in nature and within organizations. As new technologies, ideas, and methods emerge, adapting and evolving is essential to stay relevant, competitive, and efficient.

However, the change process is often riddled with resistance, inertia, and uncertainty, as individuals and organizations stubbornly cling to familiar and established ways of doing things, even when they're no longer effective or sustainable. The challenge is encouraging leadership, peers, and customers to embrace change as an opportunity for growth and renewal while addressing potential risks and obstacles.

The foundation of this concept lies in acknowledging that change is a constant and unavoidable part of life, both in nature and within organizations. As new technologies, ideas, and methods emerge, adapting and evolving is essential to stay relevant, competitive, and efficient. However, the change process is often riddled with resistance, inertia, and uncertainty, as individuals and organizations stubbornly cling to familiar and established ways of doing things, even when they're no longer effective or sustainable. The challenge is encouraging leadership, peers, and customers to embrace change as an opportunity for growth and renewal while addressing potential risks and obstacles.

NOTE: Be prepared that the people mentioned here may feel that the CRM is a threat to either them personally, their departments, or the system(s) they may have been personally a part of creating some time ago.

WATCH **out for this** - this is a very real and common problem and it causes "hard-hitting stumbles" frequently.

Which systems should be sunset?

WHEN EMBARKING on implementing a CRM system, evaluating and sunset specific systems that may no longer efficiently serve your organization's needs is crucial. As a seasoned professional in the field, I cannot stress enough the importance of this step. Through my years of hands-on work, I have learned the hard way that neglecting to address outdated systems can lead to countless headaches and hinder your CRM success.

Allow me to guide you through this process and shed light on some shared systems that should be carefully considered for sunset:

LEGACY CRMS; If you currently have an outdated CRM, it is high time to assess its relevance to your current requirements. Take a close look at its functionality, user experience, and integration capabilities. Ask yourself if it aligns with your business objectives and genuinely empowers your team. If the legacy system falls short in any of these areas, it is a strong candidate for sunset. Don't let sentimentality cloud your judgment; remember, the goal is to enhance efficiency and effectiveness.

Spreadsheets; Ah, spreadsheets. The trusty makeshift solution that many organizations have relied on for tracking customer data. While they may have served their purpose somewhat, they lack robust CRM depth and integration capabilities. Transitioning from spreadsheets to a proper CRM can work wonders for your organization. You'll experience improved data accuracy, seamless team collaboration, and enhanced operational efficiency. Leave behind the days of manually updating cells and sharing files. Embrace the power of a CRM that can centralize your customer data and provide real-time insights.

Other Data Silos; Take a moment to identify any other systems or databases within your organization that store customer information independently. These independent data silos create barriers to efficient cross-functional collaboration. They can lead to data redundancies, communication gaps, and a fragmented view of your customers. To truly unlock the full potential of your CRM, it is vital to evaluate these data silos and sunset them if necessary. By consolidating all customer information within the CRM, you can streamline operations, foster collaboration, and gain a unified and holistic view of your customers.

By evaluating and sunset these systems, you can lay a solid foundation for your CRM implementation. This proactive approach will set you up for success by ensuring that the CRM meets your organization's specific needs and effectively supports your business processes.

However, I must caution you that this process should not be taken lightly. It requires careful planning, thorough evaluation, and strategic decision-making. Rushing through it or neglecting it altogether can result in costly inefficiencies and frustrations down the road.

Let this be a turning point for your organization—a moment where you lay the groundwork for CRM success by bidding farewell to systems that no longer serve you and embracing the transformative capabilities of a robust CRM. The journey will be worth it, and your organization will be rewarded by the benefits of increased efficiency, streamlined operations, and empowered teams.

How can I determine which systems should be sunset?

MEASURING the success of a CRM system is an essential task for leaders, project teams, and administrators. As part of this process, evaluating existing systems for potential sunset becomes necessary. However, this task should not be taken lightly. It requires a thorough assessment of functionality and compatibility with the new CRM to ensure a seamless transition. This guide will determine which systems should be sunset and provide valuable insights based on years of hands-on experience in the field.

Functionality Assessment; The first step in evaluating systems for sunset is to assess their functionality. Begin by identifying potential duplicate functionality that may overlap with your new CRM. Look for areas of redundancy and determine if the functionality provided by the system is essential. Analyze whether these functionalities can be effectively transitioned and integrated into the new CRM. Assessing the necessity and feasibility of migrating functionality will help make informed decisions regarding the sunset process.

Data Evaluation; Understanding the data stored within each system is critical to determining its future. Review and analyze the quality, accuracy, and completeness of the data housed in the system being considered for sunset. This evaluation is essential to ensure that the data can be seamlessly migrated to the new CRM without compromising data integrity. Poor data quality or incomplete datasets can lead to errors, inaccuracies, and information gaps, ultimately impacting the effectiveness of the new CRM. Take the time to thoroughly analyze the data and consider any cleansing or transformation processes needed before migration.

Integration Analysis; An often-overlooked aspect of system sunset is the analysis of integration capabilities. Review the existing systems and evaluate their ability to integrate with your new CRM seamlessly. Examine the technical infrastructure and determine if any significant challenges or limitations exist concerning data migration and process transition. Integration capabilities are crucial to ensure a smooth transition, as a lack of proper integration can result in data inconsistencies and process gaps. Consult IT experts to analyze integration challenges and plan for technical solutions thoroughly.

Stakeholder Engagement; The success of any significant system migration or sunset process depends heavily on stakeholder engagement and communication. Throughout the evaluation process, involve key stakeholders, including executives, department heads, and end-users. Seek their input and insights to ensure that important considerations and requirements are considered. Discuss the implications of system sunset and address any concerns or questions they may have. Open lines of communication will foster a sense of ownership and collaboration, ultimately leading to a smoother transition and successful implementation of the new CRM.

Cost-Benefit Analysis; In addition to the functional and technical considerations, conducting a thorough cost-benefit analysis is essential before deciding on system sunset. Evaluate the financial implications of maintaining the existing system and migrating to the new CRM. Consider factors such as licensing fees, support costs, potential productivity gains, and long-term savings. Compare these costs against the anticipated benefits of implementing the new CRM, including increased efficiency, improved customer service, and enhanced data visibility. This analysis will help determine both options' overall value and return on investment, aiding in decision-making regarding system sunset.

Evaluating and selecting systems for sunset is a vital step in measuring the success of a CRM system. It requires careful assessment of functionality, data quality, integration capabilities, stakeholder engagement, and financial considerations. By following the steps outlined in this guide, leaders, project teams, and administrators can navigate the challenges of system sunset and ensure a smooth transition to the new CRM. Remember, thorough due diligence and expert guidance are essential to make informed decisions and achieve optimal outcomes. By approaching system sunset with caution and wisdom, organizations can maximize the full potential of their CRM system and measure success in a meaningful and impactful way.

What are the risks of not sunsetting other systems during CRM implementation?

IN THE COMPLEX CRM implementation landscape, failing to sunset other systems can introduce substantial risks to your organization. Administrators must understand the potential pitfalls and embrace adaptability to ensure the success of their CRM initiatives. This comprehensive guide will delve deeper into the importance of system sunset and its impact on data integrity, operational costs, and overall efficiency. Drawing from years of hands-on experience, we will provide valuable insights to help administrators navigate the challenges and make informed decisions.

Data Inconsistencies And Errors; Maintaining multiple systems simultaneously, especially those with duplicate or conflicting data, can have severe consequences for your organization. The presence of inconsistencies, inaccuracies, and errors can severely hinder data-driven decision-making and compromise your customers' trust in your organization. Imagine the frustration and disappointment a customer experiences when their information is inconsistent across different systems, leading to disjointed customer interactions and a fragmented view of their relationship with your organization. As an administrator, it is vital to recognize the impact of data inconsistencies and errors and take proactive steps to mitigate them through system sunset.

Increased Costs; Operating and managing multiple systems concurrently can trigger a significant financial burden for your organization. The costs associated with licensing, support, and resources required for upkeep can quickly escalate, especially when those systems no longer align with your organization's evolving needs. Such financial strain impedes your ability to invest in vital areas that drive customer satisfaction and operational excellence. Administrators inadvertently contribute to a multifaceted financial drain that compromises short-term profitability and long-term sustainability by failing to sunset other systems. It is in the best interest of administrators to conduct a thorough cost-benefit analysis and evaluate the value of maintaining outdated systems about the benefits of migrating to an integrated CRM platform.

Reduced Efficiency; Efficiency is the lifeblood of any successful organization. A failure to consolidate systems during CRM implementation deprives your organization of opportunities for streamlined processes, automation, and improved collaboration. When different departments operate in silos, hampered by incompatible systems, the flow of critical customer information becomes fragmented. Essential details and insights remain compartmentalized, resulting in missed cross-functional sharing and collaboration opportunities. The lack of a unified platform hampers efficiency; it impedes the organization's ability to deliver personalized and seamless customer experiences. Administrators must recognize that embracing system sunset unlocks the potential for enhanced efficiency and information sharing throughout the organization.

REPEATED WARNING: Not sunsetting all legacy systems and processes that the CRM is now performing is the silent kiss of death to the CRM - **My personal mantra**

TO MITIGATE risks and maximize efficiency, administrators must adopt a strategic and proactive approach to navigate the system sunset process successfully.

Here are some critical steps to take:

Stakeholder Engagement; Engaging all relevant stakeholders is critical to the success of system sunset. Collaborate with executives, department heads, and end-users to solicit their input and address any concerns or questions they may have. By involving key stakeholders throughout the evaluation and sunset process, administrators can foster ownership and collective responsibility, making the transition more seamless.

Functionality Assessment; Evaluate the functionality of each system targeted for sunset. Identify any duplicate functionalities that may overlap with the new CRM platform. Assess the necessity and feasibility of migrating essential functionalities to the new system. It is crucial to consider the organization's short-term and long-term needs to ensure a smooth transition without compromising critical capabilities.

Data Auditing and Migration; A comprehensive data audit is essential to understand the quality, accuracy, and completeness of the data stored in each system. Administrators must evaluate if data can be seamlessly migrated to the new CRM without compromising data integrity. Consider investing in data cleansing and transformation processes to ensure that migrated data is accurate, consistent, and reliable. Failing to pay adequate attention to data auditing and migration can result in costly errors and erode customer trust.

Integration Analysis; Evaluate the integration capabilities of existing systems. Determine if they can seamlessly integrate with the new CRM or if significant technical challenges exist. Integration capabilities are crucial for a smooth transition of data and processes. In collaboration with IT experts, assess the technical infrastructure and plan for necessary solutions to overcome any integration challenges that may arise during system sunset.

Proactive Change Management; Managing change effectively is vital for successful system sunset. Administrators should establish clear communication channels to update stakeholders on the progress and impact of the sunset process. Provide training and support to end-users to ensure a smooth transition and minimize disruption to their daily work. By proactively managing change, administrators can mitigate resistance and build enthusiasm for the new CRM, enhancing adoption rates and overall success.

. . .

IN THE EVER-EVOLVING landscape of CRM implementation, administrators must recognize the importance of system sunset. Failing to address the need for consolidation and adaptability can have severe consequences for data integrity, operational costs, and overall efficiency. Administrators can mitigate risks and maximize the benefits of system sunset by thoroughly evaluating existing systems, engaging stakeholders, and strategically planning the sunset process. Successful CRM implementation is not solely about technology; it requires thoughtful leadership, proactive decision-making, and a commitment to continual improvement. Embrace the wisdom from our experiences to chart a path toward CRM success, empowered by a unified and efficient system.

What are some common challenges with sunsetting other systems during CRM implementation?

ONE OF THE primary challenges that project teams and administrators may face when sunsetting other systems is resistance to change. It is common for people within the organization to resist letting go of familiar systems, especially if they have been using them for a long time.

Overcoming this resistance requires effective change management strategies, clear communication, and emphasizing the benefits of the new CRM system. It is important to engage stakeholders early in the process, involving them in decision-making and addressing their concerns to gain buy-in and support. By adequately communicating the advantages of the new CRM system, such as improved efficiencies, streamlined processes, enhanced data analytics, and better customer insights, project teams, and administrators can help alleviate resistance and foster a sense of excitement and acceptance among users.

Another challenge that can arise when sunsetting systems during CRM implementation is a lack of resources or expertise. The sunsetting process requires dedicated resources, technical expertise, and proper training to ensure a smooth transition. Insufficient resources or a lack of technical knowledge can impede the successful sunset of systems and hinder the adoption of the new CRM system. To overcome this challenge, project teams and administrators must thoroughly assess the resources available to them, including the availability of skilled personnel and the adequacy of training programs. This assessment will help identify gaps in resources or expertise, allowing stakeholders to address these issues proactively. It may involve bringing in external consultants or experts to provide additional support or investing in training programs to up-skill the existing team. Project teams and administrators can minimize potential roadblocks and facilitate a smooth transition by ensuring the right resources and expertise are in place.

Alongside resistance to change and a lack of resources or expertise, the risk of data loss or corruption during the sunset of systems is a significant concern that project teams and administrators must mitigate. Improper data handling during the sunset process can lead to data loss or corruption, seriously affecting the organization. To minimize this risk, it is crucial to establish proper backup procedures before beginning the sunset process. Backing up all relevant data ensures that unforeseen issues or accidental data loss can be easily rectified. Additionally, project teams and administrators should validate the integrity of the data before, during, and after the sunset process. Rigorous testing should ensure that data remains intact and accurate throughout the transition. This may involve running parallel tests or conducting staged rollouts to confirm that data is correctly migrated to the new CRM system without any loss or corruption. By incorporating these measures, project teams and administrators can significantly reduce the risk of data loss or corruption and maintain the integrity of their information.

Sunsetting other systems during CRM implementation can be challenging for project teams and administrators. Resistance to change, a lack of resources or expertise, and the risk of data loss or corruption are common obstacles that must be addressed. However, by being adaptable and implementing effective change management strategies, ensuring the availability of necessary resources and expertise, and mitigating the risk of data loss or corruption, project teams, and administrators can successfully navigate these challenges and achieve a smooth transition to the new CRM system. Proactive planning, stakeholder engagement, and a rigorous approach to data handling are vital to ensuring a successful sunset of systems and maximizing the benefits of the new CRM system for the organization.

How can I ensure that data is correctly migrated from other systems to the new CRM?

AS A SEASONED professional in the field of CRM, it is essential to understand the importance of data migration during implementation and adaptability's role in ensuring a successful transition. Data migration is an extremely complex process that requires meticulous planning, attention to detail, and the ability to adapt to evolving circumstances.

Developing A Comprehensive Data Migration Plan; The data migration process is intricate and multifaceted, and it is critical to approach it with a well-developed plan. Administrators must create a comprehensive program that outlines the data migration strategy, timeline, and responsibilities. A clear and concise plan will ensure that all stakeholders know the migration process and their role in it. It is also crucial to define the scope of the migration and set realistic goals to avoid any delays or setbacks throughout the process.

Ensuring Compatibility Of The New CRM; One of the most common pitfalls during data migration is the compatibility issue between the old and the new systems. Ensuring that the new CRM has the necessary capabilities to handle the migrated data is essential. Administrators must assess data formats, structures, and potential transformations required for successful migration. In addition, it is crucial to evaluate all the third-party applications involved in the process and ensure they are compatible with the new system.

Thorough Testing Of The Migration Process; Before fully transitioning to the new CRM, testing the migration process thoroughly is vital. Data consistency and accuracy must be validated, and testing must be done on various scenarios to catch any exceptions. Continuous testing is critical during the migration process. This can not only help avoid data inconsistencies but also provide essential insights and feedback on the overall efficiency of the migration process.

Adaptability During The Migration Process; While the steps outlined above are critical to ensuring a successful data migration process during CRM implementation, adaptability is equally essential. Administrators must be willing to adapt to new developments during the process. The migration process will not be perfect, and unforeseen roadblocks or challenges may arise. However, addressing these challenges can make all the difference in ensuring a successful data migration process.

Adaptability can also mean adapting to new technologies that may enhance the CRM system's overall functionality. Administrators must continually evaluate new technologies and enhancements that may improve the CRM system's overall efficiency and customer experience. A willingness to adapt to new technological developments is vital, allowing organizations to stay competitive in an ever-evolving business landscape.

Data migration is a complex process, and several challenges may arise during the implementation. Here are some of the challenges administrators face, along with potential solutions:

Data Integrity; Maintaining data integrity during the migration process is critical. Accuracy, completeness, and consistency of the data must be ensured. To overcome this challenge, administrators must conduct a data audit to identify duplicate data, incomplete data, inconsistencies, and errors.

Business Continuity; The migration process can cause business disruptions. Administrators must plan for temporary business process changes, such as halting transactional activity or redirecting customers to alternative channels.

Technical Challenges; Technological challenges, such as data incompatibility, third-party applications compatibility, and infrastructure, can hinder the migration process. Collaborating with IT experts to assess technical infrastructure and plan for necessary solutions to overcome any integration challenges that may arise during data migration is crucial.

Data Security; Ensuring data security during the migration process is critical. Administrators must implement rigorous

. . .

DATA MIGRATION during CRM implementation is vital to maintaining data consistency, accuracy, and integrity. Administrators must develop a comprehensive data migration plan, ensure compatibility with the new system, and thoroughly test the migration process to minimize all risks. However, even with a well-crafted plan, unforeseen challenges may arise, and administrators must remain adaptable to new developments. Evaluating new technologies is crucial to stay competitive in an ever-evolving business landscape. By adhering to these best practices, administrators can ensure a successful data migration process during CRM implementation.

Ensuring a smooth transition when sunsetting other systems during CRM implementation?

To ensure a smooth transition when sunsetting other systems, consider the following:

Transparency To Everyone; Develop a comprehensive plan with clear timelines and milestones. Share this plan with all stakeholders and ensure they understand the purpose and benefits of the new CRM.

Involve Key Stakeholders; Engage critical stakeholders throughout the process, including representatives from each department affected by the system sunset. Involve them in decision-making and encourage their input to increase ownership and acceptance of the change.

Provide Frequent Communication And Updates; Keep all stakeholders informed of the progress, challenges, and milestones. This transparency is the foundation of trust and confidence in the process.

BY INCORPORATING THESE PRACTICES, you can foster a smooth transition, address concerns early on, and gain widespread acceptance and adoption of the new CRM.

2

Requirement of Menial, Manual Primary Data Entry

As someone who has been in the trenches of CRM implementation for years, let me share some hard-earned wisdom. While data entry is a crucial aspect of CRM, it's also one of the most tedious and uninspiring tasks. Although your staff might be brave initially, they will inevitably get bored and search for excuses not to use the CRM. Ultimately, this results in missed opportunities and a lack of system uptake, which every business owner wants to avoid. CRMs are designed to be easy and user-friendly. However, if your staff doesn't find it interesting or engaging, they may not use it as intended. Consider these tips to ensure your CRM is being used effectively:

Make it fun! If you can find ways to make the system more engaging, your staff will be more likely to use it. For example, try including games or other activities that can be played on your CRM. This could mean adding a digital scratch card every time an opportunity is closed or creating a leaderboard for salespeople to see who has completed the most opportunities this month.

Create a clear purpose for your CRM, so your staff understand why they need to use it. A lack of understanding can result in frustration and a poor attitude toward the system, ultimately leading to fewer sales opportunities identified by your team.

Keep It Simple -The best thing you can do for your CRM is keep it simple and easy to use. By doing so, you'll ensure that staff can follow the process without any issues.

THE COMBINATION of concepts tries to resolve the common problem when budgeting for a CRM implementation project does not include the transition of inherited data. When companies ask their existing employees to take on extra work in exchange for recognition that is not commensurate with the required effort, it's no wonder that employees fall by the wayside. It's not that these people are lazy; they're overburdened, and they are bound to lose motivation eventually.

Despite the apparent convenience of data entry, it is a labor-intensive and time-consuming process. Verifying and combining data from multiple sources, complying with new classification laws, and undertaking data designation require significant effort and expertise. The higher the labor intensity and scope of this job, the greater the potential for far-reaching improvements in CRM processes. As such, it's essential to allocate adequate resources in terms of budget and personnel to ensure a smooth transition.

While data entry may seem like a small part of CRM implementation, it is crucial. This task can quickly hinder CRM implementation's success without adequate planning and resources. To avoid this potential pitfall, it's essential to recognize the value of data entry work, allocate the necessary resources, and do everything possible to make the process seamless and rewarding for staff. Doing so can give staff the tools to embrace CRM's powerfully.

3

Reducing Manual Effort

Bright Idea : Keep the Slice Of Pie Smaller

I always remind people of this fact. When developing a new process, in this case, something embedded into a CRM - the total new manual effort(s) of people, team, a staff class or role, division or larger must, at worst, be the same as before. Still, ideally, it should be noticeably less. If it is equal, then it must be able to show notably better outputs to both the person themselves, the department, and the company.

IF THE SLICE of pie is greater in size (meaning more manual input/effort), you need to stop and fix this issue before proceeding.

IT IS imperative to understand that the success of any CRM effort hinges on addressing issues swiftly and decisively.

This is not just a guideline but an undeniable truth learned through hard-won experience. Whether the problem arises before deployment or after implementation, the key is to take immediate action and prioritize resolution.

In the **pre-deployment stage**, addressing any issues promptly and discreetly is crucial. Act swiftly to rectify the situation, ensuring the necessary fixes are made without drawing unnecessary attention. This approach allows you to mitigate potential risks and maintain confidence in the CRM solution.

On the other hand, if issues arise **post-implementation**, it is essential to handle them transparently and promptly. Publicly acknowledging the problem demonstrates accountability and reassures stakeholders that the organization takes their concerns seriously. Swiftly resolving the issues is of utmost importance, but equally critical is openly communicating the steps taken to rectify the situation. You can regain trust and reaffirm your commitment to delivering a top-tier CRM experience by providing examples of how the resolution was achieved.

It is crucial to approach any CRM setbacks with urgency and care. <u>Whether in the pre-deployment phase or dealing with post-implementation challenges, addressing issues promptly and transparently is the key to success</u>. Adhering to this golden rule can mitigate risks and position your CRM efforts for long-term success.

Old Process (Pre CRM)

Baseline

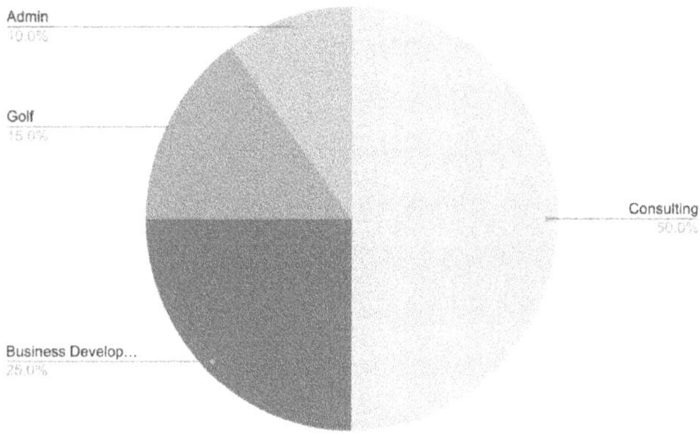

This is a demonstration of fictional data to represent a point. Pay attention to the "Admin" time/effort in particular over the next few slides.

New CRM (Reduced Manual, Better Outputs)

By Far The Best Scenario

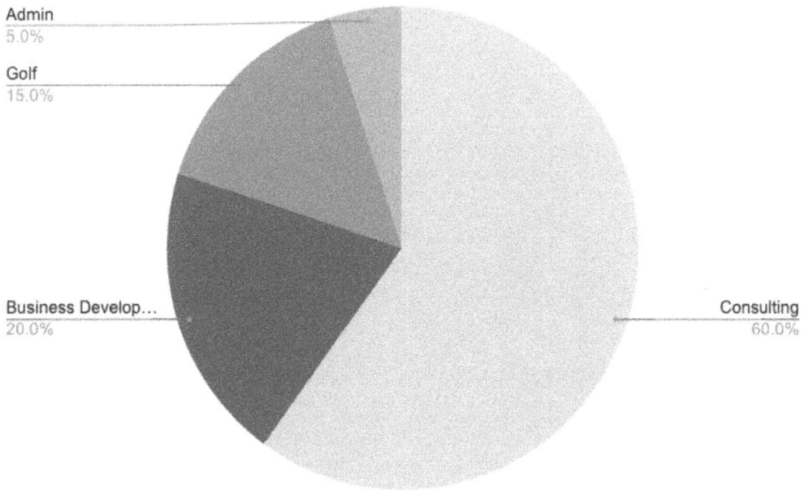

By optimizing your processes, there is the overall reduction of manual effort required, freeing up valuable time. Not only that, but the outputs have also seen a significant improvement. With enhanced reporting capabilities, deeper insights, and a wealth of knowledge available.

New CRM (Same Manual Effort, Better Output)

Ok, but not ideal

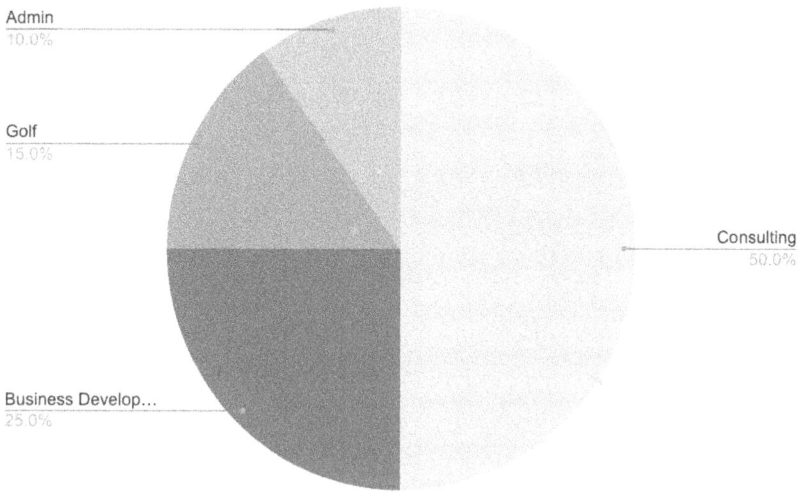

While keeping the workload the same may seem acceptable (and just squeaks by as being so), thinking critically about the bigger picture is essential. If the workload remains unchanged, users may perceive that nothing has improved, despite the enhanced outputs to their leadership. As seasoned professionals, I understand the importance of achieving results and being mindful of user perception. To this end, it's vital to reassess workflows and optimize processes to decrease workload while maintaining or enhancing output levels. By doing so, we can ensure that users positively perceive the changes made, leading to greater adoption and overall success. Remember that perception matters, and strive to make meaningful improvements that benefit the end-users and drive sustainable success.

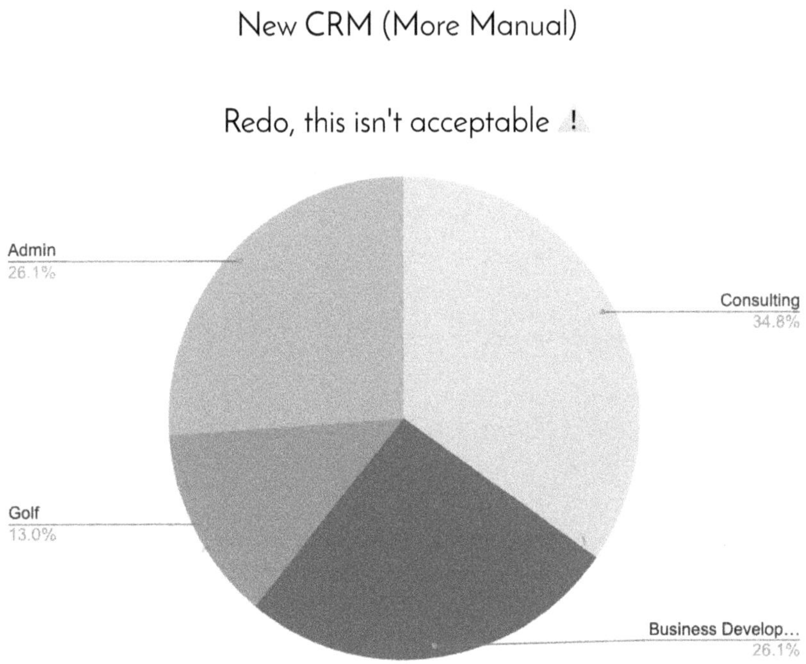

It's crucial to recognize and address the warning signs and act swiftly when the administrative effort increases. This spike in effort indicates that something may have been overlooked in the overall process, posing a significant risk to the success of your CRM system. Addressing the issue promptly is paramount. You're setting yourself up for system failure and potential user revolt by ignoring it. The negative fallout can extend beyond the CRM itself, leading people to question the system and your leadership. I've seen this scenario happen before, and you don't want to experience it. Be proactive, identify the root cause, and immediately implement necessary remedies. Protect the integrity of your CRM system, maintain user satisfaction, and safeguard your reputation as a capable leader in the process. Trust me, addressing administrative issues promptly will save you from a cascade of repercussions.

Imagine This:

YOU NEED to input the same data into an accounting information system, a service system, and who knows where else after you've entered it into your CRM. The mere thought of such a task makes even the most dedicated employee shudder. Let's face it; no rational person would find this work appealing, assuming that all the employees in your company are indeed reasonable. While most of them might grudgingly accept their fate and manually enter the same data into multiple systems, there's no guarantee that they'll have any motivation left to review the data entered and check for possible mistakes. The more employees working on the same data entry tasks, the higher the chance of disputes, repetitions due to miscommunications, missed documents, and incorrect data entry. This unfortunate state of affairs can arise because, at the time of deployment, employees were acutely aware of how this transition would chip away at their will to live (at your company). It's a grim picture, to be sure, but it's important to remember that deep integration of multiple information systems is a complex engineering challenge. And the budget allocated to this transition isn't always justified.

To reduce employee resistance to change and use of the system, conducting a business process review early in the project is crucial. This review will help determine whether workers must enter the same information into several systems due to job responsibilities. Identifying and addressing potential pain points proactively can help ensure a smoother transition and avoid the dreaded employee revolt. When various workers enter the data, it's more likely they won't experience any progressive deterioration in their morale or productivity. However, the likelihood of making a mistake increases significantly. If only one person enters the data, they'll be unhappy with the new work method, and the possibility of error will remain high. It's a delicate balancing act that requires careful consideration and planning. You could estimate the cost of unreliable data input risks if you chose not to integrate systems. When you're facing significant financial losses resulting from inconsistent data, the issue of convenience or inconvenience for your employees becomes less powerful. But, it's essential to strike a balance between financial prudence and employee satisfaction, as both factors play a critical part in the long-term success of your organization.

Now, let's delve deeper into the challenges and potential solutions associated with data entry and system integration. First and foremost, it's essential to consider the human element. Employees are, after all, the lifeblood of any organization, and their well-being and satisfaction are critical to its success. You can help create a better work environment by acknowledging their concerns and working to address them.

A critical aspect to consider is the training and support provided to employees during the transition to new systems and processes. By offering comprehensive training programs and ongoing support, you can help employees feel confident and competent in using the latest methods effectively. This can help reduce resistance to change and promote a positive attitude toward the transition.

It's also worth considering the potential benefits of adopting a phased approach to system integration. By gradually introducing new systems and processes, you can help employees acclimate to the changes more effectively and minimize the risk of disruption and dissatisfaction. This approach can also help to identify and remedy any issues or challenges that may arise during the transition, allowing for more effective problem-solving and adaptation.

Communication is another critical factor in the successful integration of multiple information systems. By keeping employees informed about the reasons for the transition, the expected benefits, and the timeline for implementation, you can help to alleviate concerns and create a more positive attitude towards the change. Open and transparent communication can also help identify and address any misconceptions or misunderstandings during the transition, ensuring everyone is working towards the same goals.

In addition to communication, collaboration is vital to successfully integrating multiple information systems. Involving employees in the decision-making process and soliciting their input and feedback ensures the transition is tailored to your organization's unique needs and requirements. This collaborative approach can also foster a sense of ownership and commitment among employees, promoting a more positive attitude towards the change and a greater willingness to embrace new systems and processes.

Worthy of critical consideration is the potential impact of system integration on your organization's overall efficiency and productivity. Streamlining and automating data entry processes can reduce the time and effort required to complete the tasks, ultimately boosting productivity and freeing up employees to focus on more value-added activities. Furthermore, by improving data accuracy and reducing the likelihood of errors, you can enhance your organization's information's overall quality and reliability, leading to better decision-making and improved business outcomes.

The successful integration of multiple information systems is complex and challenging, requiring a delicate balance between financial prudence, employee satisfaction, and technical feasibility. By adopting a proactive and collaborative approach, involving employees in the decision-making process, and providing comprehensive training and support, you can help to ensure a smoother and more successful transition. Additionally, by implementing automation tools and technologies, you can streamline data entry processes, reduce employee burden, and improve efficiency and productivity. Ultimately, the key to success lies in striking the right balance between these various factors, ensuring that your organization is well-equipped to navigate the challenges and opportunities of the modern business landscape.

4

Multiple Manual Entry Points? Seriously ?

Regarding CRM implementation, one issue that often discourages users is the requirement for manual primary data entry. Users may perceive this task as time-consuming and tedious, leading to frustration and low adoption rates. Additionally, manual data entry can result in incomplete or inaccurate data, which hampers the effectiveness of the CRM system. As a seasoned CRM professional, I've encountered this challenge and learned valuable lessons on addressing it effectively.

To truly optimize your CRM system and address the challenges, it is essential to reduce manual input and improve data quality. This aspect often gets overlooked, but it significantly impacts overall efficiency and personnel morale.

To begin, it is crucial to encourage other leaders within the organization to fully embrace the use of the CRM system for data collection. This can be achieved by integrating the CRM with existing tools and procedures. By doing so, you can automate manual tasks and develop efficient data entry techniques. This reduces duplicate efforts and enables the CRM to capture relevant data automatically. Streamlining the data entry process and minimizing the need for manual input can alleviate the burden on users, improving their experience and freeing up their time for more valuable tasks.

In addition to reducing manual input, improving data quality is equally important. Inferior data quality can lead to erroneous insights, incorrect decision-making, wasted resources, and people believing the entire system cannot be trusted. One practical step is to carefully examine all the required "fields" within the CRM system and identify where the data already resides within the organization. This exercise helps uncover redundancies and unnecessary entry points, which can be streamlined or eliminated. By reducing the number of entry points and ensuring that data is captured at the appropriate stage, you can simplify the data entry process and enhance data quality.

ADDRESSING these data-related challenges is critical for optimizing CRM efficiency and the morale of the personnel involved. Manual input and redundant processes can often lead to employee frustration and burnout. By streamlining data entry and reducing duplicate efforts, you demonstrate a commitment to improving working conditions and alleviating the burden placed on your team.

HOWEVER, I must caution that achieving these improvements may require careful planning, collaboration, and thorough testing. Integrating different tools and systems, automating tasks, and developing efficient data entry techniques can be complex. Conducting due diligence and involving experienced personnel is crucial to ensure a seamless and successful implementation.

By reducing manual input to a single entry point, improving data quality, and considering personnel morale, you can unlock the true potential of your CRM system and create a more productive and satisfying work environment for your team.

IDENTIFYING MANUAL INPUT POINTS:

Conduct a thorough audit; of all system data entry points and processes. Identify where manual input is required and the frequency of entry.

Analyze data points; to determine which are necessary and which might be redundant. Consider if specific fields can be eliminated or merged to reduce manual input.

Consult with different departments and user groups; to understand their data collection needs and the processes they follow. Determine if there are opportunities for consolidation or automation that can streamline the data entry process.

Utilize analytics; to identify the data fields that are most frequently entered and their associated error rates. This can provide valuable insights into where automation would have the most significant impact.

REDUCING MANUAL INPUT POINTS:

Integrate existing systems and tools with the CRM system; to enable automatic data capture. This could include integrating with email or social media platforms or enabling automatic syncing with commonly used applications.

Develop efficient data entry techniques; such as pre-populating fields with frequently used data or allowing users to import data from Excel spreadsheets.

Encourage other (all) departments to fully embrace; the use of the CRM system by providing training and support. This can help minimize data redundancies and eliminate the need for duplicate data entry.

Automate more complex data entry tasks; utilize advanced techniques such as machine learning or robotic process automation (RPA).

When attempting to reduce manual input points, it's essential to approach the process strategically and collaboratively. Remember that not all manual inputs can or should be eliminated, and it's crucial to consult with different departments and user groups to determine what is necessary for accurate data collection. Finally, it's essential to thoroughly test any changes or enhancements to ensure they have the intended impact and do not create new data integrity issues.

Promoting these improvements and highlighting the reduction in manual input places is very important. Communicating the benefits of these changes to users and other stakeholders is crucial in garnering support and buy-in. Emphasize how these improvements will save time, reduce errors, and provide more accurate data for better decision-making. Through targeted communication, training, and documentation, users can understand the value of the changes and feel confident in their ability to leverage the CRM effectively. Addressing the requirement of menial, manual primary data entry is essential for ensuring a successful CRM implementation. We can streamline processes and alleviate user burdens by encouraging other leaders to allow the CRM to collect information, integrate systems, automate tasks, and provide training on efficient data entry techniques. Improving existing processes, reducing entry points, and promoting these improvements are vital steps to overcome user resistance and increase adoption rates. Remember, even small changes in data entry processes can significantly impact user satisfaction and the overall success of the CRM implementation.

Adding The Same Stuff Twice

I HAVE ENCOUNTERED the issue of redundant data entry across multiple software systems and fully understand its negative consequences. Duplicate data entry can lead to user frustration, low adoption rates, and increased error probability, undermining the system's effectiveness. In this book, I will share my hard-earned wisdom and highlight the importance of addressing this issue to enhance efficiency and user experience.

One of the primary challenges associated with duplicate data entry is the frustration and inconvenience this practice creates for users. Entering the same data into multiple software systems is time-consuming and can decrease productivity and negatively impact the entire team's efficiency. By eliminating this duplicate data entry burden, we can free up users' time and enable them to focus on more valuable tasks.

Recognizing the need for improvement is crucial to achieving successful CRM implementation. By addressing this issue and highlighting the potential risks associated with errors caused by duplicate data entry, we can promote the necessary changes to streamline workflows, ensure better data quality, and improve overall user experience.

We know that one of the key benefits of implementing a CRM system is streamlining data entry processes. By reducing manual data entry and integrating CRM with other software systems, organizations can unlock significant efficiency and user experience improvements. While these benefits can be felt broadly across the organization, certain departments are almost certain to experience improved processes and boosted morale from such workflow improvements.

Let's take a closer look at some specific departments that are likely to see notable improvements:

. . .

FINANCE AND ACCOUNTING:

Streamlined Sales Order Processing; By integrating the CRM system with the finance and accounting software, sales orders can flow seamlessly into the system, eliminating the need for manual entry and reducing the risk of errors. This saves time, improves accuracy, and enhances cash flow management.

Enhanced Financial Reporting; With the CRM system capturing customer and sales data in real-time, finance and accounting departments can generate more accurate financial reports and gain deeper insights into customer behavior, enabling informed decision-making and strategic financial planning.

Customer Service:

Efficient Ticket Management; When customer service representatives can access a centralized CRM system, they can easily track and manage customer inquiries, complaints, and support tickets. This streamlines response times improves customer satisfaction, and boosts the overall efficiency of the customer service department.

Personalized Customer Interactions; By having a complete view of customer interactions, preferences, and purchase history, customer service representatives can deliver personalized and tailored solutions to customers, strengthening customer relationships and loyalty.

Sales:

Improved Lead Management; The CRM system enables sales teams to track and manage leads effectively throughout the entire sales cycle. The sales department can optimize its lead management processes, increase conversion rates, and drive revenue growth by automating lead assignments, nurturing workflows, and follow-up reminders.

Enhanced Sales Forecasting; With accurate and up-to-date data residing in the CRM system, sales managers can generate more reliable sales forecasts. This aids in resource allocation, goal setting, and overall sales performance management.

Other departments may also benefit from reduced manual data entry and streamlined processes. For example:

Marketing; Improved customer data accessibility and segmentation capabilities in the CRM system enable marketers to deliver more targeted and personalized campaigns, resulting in higher engagement and conversion rates.

Operations; Integrating the CRM system with inventory management and supply chain software facilitates better demand planning, inventory control, and order fulfillment, improving operational efficiency.

Executive Management; Access to real-time business insights and analytics from the CRM system empowers executives to make informed strategic decisions, monitor performance, and identify growth opportunities.

By implementing a CRM system and emphasizing the value of data entry, organizations can reap these department-specific benefits and achieve broader improvements such as improved customer insights, enhanced decision-making, and better business outcomes. Remember to stress the importance of due diligence and highlight potential pitfalls in a cautionary tone, ensuring that the implementation process is approached thoughtfully and diligently.

So leverage your hard-earned wisdom and experience to guide your organizations through potential challenges and pave the way for a successful CRM implementation that positively impacts all departments involved.

5

Case Study: Internal Team "Users"

Scene:

The marketing department had access to a feature-rich CRM platform. This platform boasted robust internal ticketing and IP storage workflows, providing a solid foundation for their operations. Not only that, but they also had large installations of third-party campaign tools such as Marketo, HubSpot, and Pardot. It is the perfect setup, offering the best of both worlds regarding functionality and integrations.

Incident:

One of the most frequent challenges I have faced is when a company's internal departments, such as the marketing department, want all the benefits of the CRM system without being responsible for managing any part of it - a recipe for disaster.

I once worked with a client where the marketing department desired complete control over their strategic direction and access to detailed daily reports and assignments. However, this department did not want to be involved in managing the technical platform to include even the most basic tasks. They envisioned a magical solution where everything would work flawlessly without their assistance or commitment, financially and regarding headcount.

Unfortunately, this unrealistic expectation resulted in a poorly implemented CRM system. As a result, the marketing team could not take full advantage of the tool's features, leading to finger-pointing, frustration, and complaints. The organization suffered from wasted resources and limited productivity, ultimately causing the CRM system to become more of a hindrance than an asset.

Understanding that implementing a CRM system requires effort and commitment from all parties involved is imperative. There are no shortcuts when it comes to achieving success in CRM. Every department that uses the system must have extensive involvement, including training, data entry, routine maintenance, and system updates.

Many businesses often fall into this lack of commitment and unrealistic expectations regarding CRM systems. As seasoned professionals, I advise setting realistic expectations and investing in adequate resources for the system's implementation, training, and management. Doing so can help ensure the system is efficiently and effectively used, allowing businesses to gain its full benefits and support its strategic goals.

Result:

A multi-year, finger pointing, budget grabbing **"shitshow"** that unfolded within a company's CRM, all thanks to the contradictory behavior of this internal department. Allow me to recount this experience, highlighting the importance of clear communication and collaboration when implementing a CRM.

The marketing department claimed to be "artists" who preferred to focus solely on their creative endeavors, dismissing any involvement in the technical aspects of the CRM. On the surface, it seemed like a straightforward arrangement - the marketing team would provide their needs and requirements to the CRM administrators, who would then handle the technical implementation. Easy, right? Well, not quite.

Something peculiar happened when the marketing department explained their needs to the CRM administrators. They transformed into technical experts, scrutinizing every technical detail and demanding perfection. Emergency fixes were requested, and their expectations reached unprecedented heights. Goodwill and collaboration seemed encouraging, at least for a few days.

However, as time went on, something changed. The marketing department quickly reverted to its old ways, shirking its responsibility and providing no assistance to the already understaffed admin team. The burden fell squarely on the shoulders of the CRM administrators, who were left scrambling to meet the ever-changing demands of the marketing department. The result? Chaos and frustration that persisted for years.

This cautionary tale highlights the need for consistent and clear organizational communication during CRM implementation. It is crucial for all teams involved, especially the marketing department, to fully understand their responsibilities and actively participate in the process. When departments contradict themselves and fail to provide consistent support, it can lead to misaligned expectations, delayed progress, and an overwhelming burden on the CRM admin team.

When the marketing department understands the technical aspects of the CRM and actively participates in its implementation, it can effectively communicate its needs, provide valuable input, and support the admin team. This level of collaboration will not only alleviate the burden on the admin team but also ensure that the CRM is tailored to the specific requirements of the marketing department.

Lessons Learned:

It became abundantly clear that every team, division, and business unit must step up and contribute to using a CRM. There can be no room for excuses or exceptions. We must all embrace our role as "**digital CRM corporate citizens**" to reap the benefits of a CRM.

To truly maximize the potential of a CRM implementation, active engagement in content creation and support for the system's inner workings is crucial. It's not enough to simply sign up and expect miracles to happen. We must actively participate in shaping and maintaining the CRM, ensuring that it aligns with our specific needs and objectives.

Remember, successful CRM implementation requires collaboration and involvement from all stakeholders. It's not just the responsibility of the IT department or a select few CRM experts. Regardless of their role, every team member must recognize the importance of their contributions and actively participate in the process.

By embracing our roles as "**good digital CRM corporate citizens**," we can help drive the success of the CRM and unlock its true potential. Let's collaborate, provide valuable insights, and actively shape the CRM implementation to ensure its success throughout the organization. This collective effort will pave the way for improved efficiency, streamlined processes, and enhanced customer experiences. It's up to us to make it happen.

What Should Have Been Done Differently:

While every case is unique and requires tailored approaches, in this particular situation, it would have been beneficial for the Marketing Department to have more restricted access and benefits from the CRM platform. This would have ensured they actively participated in joint support efforts to operate and advance the platform. By setting appropriate levels of involvement, the marketing department would have been more accountable and less likely to burden the admin team. This approach would have fostered a more harmonious and successful CRM implementation for the organization.

6

Case Study: Lack Of Support And Ethical Work Habits

Scene:

I once assisted a globally separated organization that decided to adopt a CRM system. With divisions operating independently and diverse workflows, processes, and practices in place, the organization thrived on its autonomy, fostered innovation, and embraced agility. However, it also led to data silos and inconsistent practices across divisions.

As we embarked on the CRM implementation journey, it became obvious that the lack of standardization and coordination posed substantial roadblocks. Each division had its own set of data fields, naming conventions, and business rules. These disparities compromised data quality, hindered effective data sharing and analysis, and impeded the organization's ability to obtain a comprehensive view of its customers.

Furthermore, resistance from individual teams within this decentralized structure posed a significant challenge. Some teams viewed the introduction of the CRM system as an additional burden or intrusion into their established ways of working. This lack of alignment and commitment from key divisions hampered the successful adoption and full utilization of the CRM system, curtailing its potential benefits.

To navigate these treacherous waters successfully, it became evident that a comprehensive change management strategy involving all divisions was imperative.

Incident:

In this organization, a recurring conflict revolves around implementing a CRM CRM system. Three distinct perspectives emerge, creating tension within the organization:

- **Division A:** This division claims to be too small in revenue to afford a CRM. They argue that investing in such a system would strain their limited resources and disrupt their operation.
- **Division B:** As a small and geographically isolated division, they argue that CRM is even more crucial for their success. With fewer existing contacts and less overall experience, they see CRM as a tool to help expand their network and build stronger relationships.
- **Division C:** This division asserts that its contribution to the CRM would be minimal. Their lack of valuable inputs justifies their primary role as information extractors. They argue that investing time and effort into adding knowledge to the system is not worthwhile for them.

Result:

When divisions operate with their own rules, processes, and objectives, it can be challenging to consolidate customer data into a unified view. This lack of cooperation can limit the effectiveness of the CRM and impact the organization's ability to leverage the benefits of centralized customer information.

It's important to approach the CRM implementation process with a comprehensive plan in place, one that **emphasizes collaboration, communication, and the importance of a unified view of customers across all divisions**. Don't let this divisional divide hinder your organization's growth potential.

To ensure success, establish clear governance mechanisms that promote collaboration and consistency in data management across all divisions. Foster a culture of communication to help keep teams connected and aligned.

Lessons Learned:

Collaboration is critical; The incident demonstrates the importance of collaboration within a globally separated organization. Instead of focusing solely on individual needs and priorities, working together towards a common goal is essential. When divisions collaborate and pool their resources, the organization benefits.

Size doesn't determine necessity; Division B highlights the misconception that smaller divisions or teams do not require a CRM. Being small and geographically removed often means having fewer contacts and less experience, making the need for CRM even more critical. By embracing CRM, small divisions can strengthen connections, make valuable insights, and facilitate growth.

Value creation goes both ways; Division C's viewpoint reveals a flawed understanding of the purpose of CRMs. CRM is not just about extracting information; it is a tool for knowledge creation and value exchange. Every division has unique insights to offer, regardless of its size or geographical location. Divisions can extract more valuable information from their peers by contributing to the system.

What Should Have Been Done Differently:

Clear communication; The organization should foster open, transparent communication channels to address concerns and misunderstandings. Divisions can better understand each other's needs and work towards mutual agreement by promoting dialogue.

Inclusive decision-making; Instead of solely relying on revenue-based justifications, decisions concerning CRM implementation should consider the long-term benefits to the entire organization. All divisions should have a voice and representation during the decision-making process.

Equal responsibility-sharing; Each division should recognize its obligation to contribute to the CRM. By actively participating and sharing valuable knowledge, divisions can benefit from the insights of others and create a more comprehensive and effective CRM ecosystem.

BY EMBRACING COLLABORATION, understanding the actual value of CRM, and prioritizing ethical behavior, organizations can overcome divisional barriers and maximize the potential of their CRMs for the benefit of all.

7

Ownership And Owners

CHILDISH BEHAVIOR

Regarding CRMs and the ownership and sharing of data and information, it's crucial to approach the concept of "ownership" with a critical lens. The term "owner" can create more problems then it intends to resolve. While having a central person responsible for coordination is essential, complex CRMs require a collective effort from individuals who have something to gain from the system's success and lose if it fails.

Rather than using the term "owner," let's reframe it as "champions." These champions play a critical role in moving efforts forward and ensuring the success of the CRM. As the central person coordinating these efforts, it's crucial to identify and empower these champions.

So, who are these champions? Let's dive into it:

Sales Representatives; Salespeople interact with customers and prospects on the front lines. The effectiveness of the CRM directly impacts them, as it streamlines their sales processes, helps them track leads, and provides insights into customer preferences. Engaging your sales representatives and ensuring they see the value of the CRM will make them powerful champions.

Marketing Professionals; Marketing teams heavily rely on

CRM data to segment customers, create targeted campaigns, and analyze the effectiveness of their efforts. Involving marketing professionals in the CRM's design and implementation will help them take ownership and become champions who can leverage the system's capabilities to drive results.

Customer Support Representatives; Customer support teams are crucial in understanding customer needs and addressing concerns. By integrating CRM with customer support processes, representatives can access valuable customer information quickly, improve response times, and provide personalized experiences. Empowering support representatives to advocate for the CRM will enhance their effectiveness and satisfaction.

Data Stewards/Administrators; Data plays a pivotal role in CRMs. Data stewards and administrators ensure data integrity, consistency, and security. By involving these individuals from the beginning, they can take ownership of data quality and collaborate with other champions to drive data governance best practices.

Executive Sponsors; Strong executive support is crucial for the success of any CRM implementation. Having executives as champions helps create a culture that values the CRM and supports continuous improvement. Executives can also clear roadblocks, allocate necessary resources to drive adoption and ensure the CRM aligns with overall business objectives.

REMEMBER, the key is identifying and engaging individuals with a stake in the CRM's success. Reframing them as champions, we emphasize their crucial role in driving the system's effectiveness. Empower these champions by involving them in decision-making, providing adequate training, and recognizing their contributions to reinforce their commitment.

Together, as a team of champions, you can foster a culture of shared responsibility, collaboration, and accountability, ultimately maximizing the value of your CRM for your organization and its customers.

8

Case Study: Performers, Champions, And Sponsors

Scene:

The excitement of a new CRM deployment can be palpable. It's a time of great anticipation as we look forward to the benefits and improvements it will bring. But with any significant change comes concerns about how the users will embrace it. As seasoned professionals in the CRM space, most know that change can be challenging, and we must approach it with caution and diligence.

The project team has undoubtedly worked months hard to ensure a smooth transition. However, it's critical to appreciate that user adoption is essential to the success of any CRM deployment. It's up to the leadership team to ensure the company is ready and willing to adopt the new system. Failure can lead to user resistance, reduced productivity, and lost opportunities.

As the leadership team gathers in a conference room, it's essential to discuss the upcoming deployment, potential challenges, and how best to support users through the transition. Open and honest discussions about the deployment's challenges and potential pitfalls are essential to success.

It's vital to approach the deployment with caution and diligence. Ensuring users are trained on the new system and support is available for their questions, and concerns is crucial. The organization must be prepared to adapt and embrace the new system fully.

As we reflect on our experience in the CRM space, we know that due diligence, thorough planning, and strong leadership are keys to success. It's essential to approach the deployment with a cautious and mentor-like tone, highlighting the importance of user adoption and providing support at all levels. With the benefit of our experience, we can steer the organization toward a successful CRM deployment and deliver the many benefits it promises.

The office is excitedly buzzing as the new CRM is being deployed. The project team has put in months of hard work to ensure a smooth transition, but there are concerns about how the users will embrace the change.

The leadership team gathers in a conference room to discuss the upcoming deployment and the need for solid support from all levels.

Incident:

Unfortunately, resistance to change can lead some users to stick to the old, soon-to-be-retired systems and processes.

Many leaders may still accepted reports and data requests submitted through outdated methods. This quickly led to problems. Data management became confusing as there were now two different ways of doing things. Redundancies and errors followed, resulting in lost time and the need for extra resources to clean up data and correct errors.

This scenario is disastrous, and as seasoned CRM professionals, we are responsible for mitigating the risk of such situations. Proper communication and training ignored vital things to ensure user adoption and successful deployment. Leaders were not trained to understand and embrace the ramifications of accepting submissions via old methods and reinforce the importance of submitting data through the new system.

Red flags were not seen fast enough to caution leaders at every opportunity about the potential pitfalls of user resistance and noncompliance with the new CRM.

Many underestimated the importance of proper mandates, communications, and training when rolling out a new CRM.

Result:

The team encountered a situation where some users' lack of strong leadership support and resistance to change proved significant obstacles in successfully deploying our new CRMs. The consequences were far-reaching and impacted the organization's ability to fully realize the benefits of increased efficiency and streamlined processes.

Teams struggled to align their work with the new system without robust leadership support. The resistance to change created a ripple effect that hindered progress and productivity. Instead of focusing on the tasks, time and effort were wasted on managing the transition. It was a frustrating and disheartening experience.

Reflecting on this challenging period, it's clear that the team could have mitigated these issues with more robust leadership support and proactive change management strategies. Engaging leaders from the beginning and enlisting their total commitment to the new system would have been crucial. Leaders could have set the tone and expectations for the entire organization by championing the benefits and driving the adoption of CRMs.

To navigate such challenges, it's essential to prioritize strong leadership support and communication throughout the CRM deployment process. Leaders must be aligned and fully committed to the transition, understanding the need for change and the potential benefits it can bring. By doing so, teams can be inspired to embrace the new CRMs, ensuring a smoother deployment and maximizing the intended efficiency gains.

Our past experiences have taught us the importance of due diligence in managing change in CRM deployments. By recognizing potential pitfalls, such as the lack of leadership support and resistance to change, we can take proactive measures to address them. In doing so, we can minimize the waste of time and effort and focus our energy on achieving the organizational goals at hand.

Let our shared experiences guide us toward solid leadership support and proactive change management strategies in future CRM deployments. By prioritizing these elements, we can steer our organizations toward greater success, fully realizing the intended benefits of increased efficiency and streamlined processes.

Lessons Learned:

Strong land vocal leadership support is crucial; For any significant change initiative, like implementing a new CRM, leadership must show unwavering support and clear communication. It's not enough for leaders to quietly endorse the change; they must actively embrace and vocally advocate for the new system.

Mandate the use of new systems and processes; Leaders must accept that "mandate" is not a four-letter word, but "you-f*cked-yourself" should be in your attempt to be a compassionate leader. To prevent any confusion and resistance, it is essential to communicate in all formats possible that the old systems and procedures will be retired and the new system must be used. This ensures a uniform approach across the organization and avoids conflicts between different versions of truth.

Minimize the effect of resistance; Resistance to change is natural, but addressing any signs of rebellion early on is essential. Users need time to adapt, and with strong leadership support and effective change management strategies, they can be guided to embrace the new system.

What Should Have Been Done Differently:

Looking back, the organization should have taken the following steps to ensure a more successful deployment of the new CRM:

Set Clear Expectations And Consequences; The leadership team should have communicated that the old systems would be retired and the new CRM must be used. They should have set clear expectations for compliance and communicated the consequences of not adhering to the new processes.

Provide Comprehensive Training And Support; User resistance may stem from fear of the unknown. The organization should have invested in extensive training and support resources to help users transition smoothly to the new system. This could include hands-on training sessions, user guides, and ongoing support from a dedicated helpdesk.

Address Concerns And Provide Feedback Channels; It's essential to create a culture of open communication and address any concerns or questions from users. Leadership should have encouraged feedback and suggestions to ensure user needs and pain points were considered during deployment.

By taking these additional steps, the organization could have minimized resistance, maximized the benefits of the new CRM, and accomplished a successful deployment within the given time constraints.

9

Case Study: Wizard Technology Skills - Improper Project Team

Scene:

Picture a small company brimming with pride in its intelligence and business acumen. Under the leadership's guidance, this belief has permeated the organization, fostering a sense of self-reliance that led them to believe they could develop and implement their CRMs without seeking external assistance. The IT department, always eager to showcase its technical prowess, eagerly approaches the challenge. On the surface, this unwavering stance and desire to save money while flaunting their skills might appear noble. However, my seasoned eye recognizes the precarious ground on which they tread.

In this scenario, it's essential to remember that CRM implementation is not simple. It requires careful orchestration of technical expertise, industry best practices, and change management strategies. While their self-confidence is admirable, attempting to tackle the entire process in-house without external guidance can lead to numerous pitfalls.

One potential challenge lies in underestimating the complexity of CRMs. The intricacies of data management, workflow automation, and integration with existing systems can easily overwhelm even the most skilled IT department. Without access to specialized CRM knowledge and experience, the company may reinvent the wheel, wasting time and resources on trial and error.

Furthermore, without external assistance, they miss out on the valuable insights and best practices that consultants and CRM vendors bring. These experts have navigated similar deployments countless times, learning from successes and failures. Their guidance can help steer an organization from common pitfalls and towards a more efficient and effective CRM.

Additionally, we must not overlook the importance of change management. Implementing a new CRM affects the entire organization, from sales and marketing teams to customer service and support. It requires a shift in processes, mindsets, and behaviors. Without a comprehensive change management strategy, the company risks encountering user resistance, decreased adoption rates, and a significant gap between the intended benefits and the actual outcomes.

While the desire to save costs and showcase internal skills is understandable, embarking on a CRM implementation without external assistance can be treacherous. Let us instead advocate for due diligence, engaging CRM experts for their specialized knowledge and harnessing their experience to navigate the complexities of CRM implementation. Doing so can bolster our chances of success and maximize our CRM's benefits to the organization.

Incident:

This frugal company embarked on a CRM implementation without seeking external guidance or expertise. Fueled by overconfidence and a belief that the challenges would be easily overcome, they forged ahead without fully grasping the intricacies of the CRM network and the potential hurdles lurking ahead.

In their eagerness to showcase their skills, the IT department took charge of developing and implementing the CRM. At the outset, they were filled with excitement and optimism. However, as time passed, problems began to rear their heads, casting a shadow over their grand plans.

One of the critical issues that emerged was the lack of a holistic understanding of the CRM ecosystem. The team underestimated the complexity of integrating the CRM with existing technologies, databases, and process change. They soon grappled with compatibility issues, data inconsistencies, and poor system performance. What initially seemed straightforward became a tangled web of technical challenges.

Moreover, their decision to bypass external expertise meant they missed invaluable insights and best practices. The abundance of resources and knowledge from CRM consultants, vendors, and solution providers could have helped them navigate the treacherous terrain more effectively. These experts have encountered many scenarios and can offer guidance on common pitfalls, system optimization, and tailored strategies for seamless integration.

As time passed, my dear colleague, the company's lack of external guidance led to frustration and setbacks. The development and implementation process was prolonged, consuming precious time and resources as problems mounted. The initial excitement within the IT department gave way to stress, and their confidence wavered.

Reflecting on this experience, it is clear that the frugal approach taken by the company, though understandable, was not without its pitfalls. Sometimes, the desire to save costs by avoiding external assistance can result in far more significant expenditures in the long run. Time spent on troubleshooting and reworking the CRM could have been better used on strategic initiatives that drive business growth.

When venturing into a CRM implementation, it is imperative to approach it with caution and due diligence. While self-reliance and cost-saving are admirable principles, they must be balanced with the need for external expertise and guidance. Engaging CRM consultants, leveraging vendor support, or partnering with experienced solution providers can help pave the way for a smoother implementation journey, minimizing challenges and maximizing the benefits of the CRM. Let us tread wisely and draw upon the lessons learned from past encounters to steer our organizations toward success in CRM implementations.

Result:

As the company's project progressed, it became painfully apparent that they had not adequately anticipated or equipped themselves for the challenges ahead. The integration process was marred by compatibility issues with pre-requisite systems, casting a shadow over the development endeavor.

These compatibility issues posed significant obstacles and hindered progress at every turn. The team faced the daunting task of ensuring seamless integration between the CRM and the existing infrastructure. However, they were ill-equipped to overcome these challenges without thoroughly understanding the intricacies. The resulting delays and setbacks exacerbated their plight, impeding the CRM's timely launch and smooth functioning.

The team's lack of proficiency in CRM functionalities and complex business processes fueled the fire. They grappled with a system they did not fully comprehend, leading to subpar performance and underutilization. The initial vision of leveraging the CRM to streamline and optimize their operations remained unfulfilled.

Consequently, disillusionment abounded among the employees. They witnessed missed opportunities slip through their fingers. These opportunities could have been seized and capitalized upon with a properly implemented and fully functional CRM. This lack of foresight and preparation left the team disheartened and demotivated as they struggled to meet their objectives and achieve their envisioned benefits.

This situation is a stark reminder of the importance of due diligence and comprehensive preparation in CRM implementations. It underscores the need to deeply understand the system's functionalities and how it aligns with complex business processes. Investing in the necessary training and acquiring the expertise needed to operate and optimize the CRM effectively is crucial to avoid falling into the same predicament.

Lessons Learned:

The company should have acknowledged the knowledge gaps before the CRM implementation began. They should have honestly assessed their abilities and identified knowledgeable staff that could have integrated with third-party vendors to avoid making the same mistakes.

Engage CRM Experts; The stakeholders and decision-makers should have sought external assistance from CRM experts or consultants who can provide guidance and support throughout the CRM implementation process. The experience and knowledge of these professionals can help navigate complex challenges and ensure a successful CRM implementation.

Invest In Training And Education; The company can evaluate available training programs or workshops and invest resources in training to build internal expertise and reduce reliance on external third-party service providers. This step should have begun at onboarding and continued through the system's deployment in all anticipated phases of future growth.

What Should Have Been Done Differently?

In my years of experience in the field of CRM implementation, one of the most important lessons I've learned is that it's always hard to know what you do not know. 🤪 This knowledge gap can lead to the negative consequences of overconfidence, which can significantly impact the success of CRM initiatives. Reflecting on past projects, I can't help but think about the steps a company should have taken to avoid these pitfalls.

First and foremost, the company should have recognized its expertise gaps. Understanding the level of expertise required for CRM implementation is crucial. Without this recognition, the organization may have developed false confidence, unaware of the potential challenges and complexities ahead. In this scenario, seeking external assistance from CRM consultancy providers with extensive knowledge of the CRM platform, architecture, and functionalities would have been wise. These experts could have provided the guidance needed to navigate the complexities of the implementation process, ultimately minimizing the risks associated with overconfidence.

In addition to seeking external guidance, investing in holistic CRM education and training should have been a priority for the company. It's not enough to simply implement a CRM and assume that employees will intuitively know how to use it. Every staff member using the implemented CRM should receive comprehensive training and education on its features and functionalities. This investment in knowledge would have empowered the team with the necessary skills to leverage the CRM effectively, maximizing its potential and minimizing the risk of underutilization.

It's important to emphasize that overconfidence in one's abilities during CRM implementation can expose an organization to various risks. Acknowledging knowledge gaps, seeking external guidance, and investing in comprehensive employee training and education are critical to mitigate these risks. By making informed decisions and aligning CRM implementation with the company's goals, organizations can set themselves up for long-term performance excellence.

The lesson learned from past experiences is clear: to avoid the negative consequences of overconfidence in CRM implementation, it's imperative to recognize expertise gaps, seek external assistance, invest in holistic CRM education and training, and make informed decisions aligned with strategic objectives. By adopting a cautious and diligent approach, organizations can confidently navigate the complexities of CRM implementation, ensuring long-term success in their CRM initiatives.

10

Case Study: Death By Steering Committee

Scene:

The room was bustling with energy as team members began to share their ideas. The walls were lined with whiteboards, covered in diagrams, and filled with written notes that showcased each team member's grand vision and ambitions for the project. It was a moment of great anticipation, and every participant was eager to share their ideas, concerns, and objectives.

While the enthusiasm was palpable, there was a degree of caution among some team members. They recognized that implementation projects can be challenging and complex and that even the best ideas can encounter unexpected obstacles. The seasoned project leader professional understood the importance of balancing ambition with a practical, realistic approach to project planning.

Incident:

In their relentless pursuit of delivering the perfect system, the organization made a well-intentioned decision to involve as many people as possible in the decision-making process. The plan was noble - to gather diverse perspectives and ensure that all stakeholders had a voice in shaping the system. However, as I have seen repeatedly, this decision had unintended consequences. Creating a mega team comprising 20 individuals, each with a strong personality and unwavering ideas of what the system should be, set the stage for conflicts and challenges.

As the project unfolded, it became evident that internal divisions plagued the team. Different factions emerged, passionately advocating for their vision of the perfect system. Collaboration and effective communication, crucial elements for successful CRM implementation, took a backseat as selfishness and egos prevailed. The team was mired in disagreements, with progress hindered by the inability to find common ground and align on shared goals.

The consequences were far-reaching. Decision-making became arduous and time-consuming as each individual fought relentlessly for their objectives. Deadlines slipped, and tensions simmered within the team. The project's momentum was stifled, and frustration grew among team members who could not make meaningful progress.

In the face of this challenge, it was clear that the team needed external guidance and support. An experienced facilitator could have helped steer the team towards common objectives, enabling them to overcome differences and develop realistic, shared goals. This guidance could have instilled a sense of unity and purpose and fostered an environment of collaborative problem-solving.

Result:

I have seen firsthand how the consequences of certain decisions can have far-reaching implications. In many cases, the cost of these decisions goes beyond financial considerations and impacts company reputation and staff morale. The results of organizational oversights and lack of coordination can be disastrous, especially regarding CRM implementation.

In the case of the organization described above, the consequences of its decision to involve too many people in the decision-making process became apparent as the CRM implementation project moved forward. With an overabundance of ideas and insufficient coordination, the resulting system was disjointed and convoluted. Conflicting functionalities were incorporated, resulting in confusion and frustration among users.

Moreover, including unnecessary features resulted in wastage of critical project resources, pushing up the overall development cost. Ultimately, the system failed to meet the organization's and users' needs, leading to disappointment and frustration. This failure significantly affected the organization's reputation, and staff morale took a hit as project participants struggled to understand what had gone wrong.

CRM implementation projects can often be complex and multifaceted, and organizations must invest adequate time, resources, and expertise to ensure they're executed successfully. Decisions should be made based on pragmatic considerations rather than being led by the desire to involve as many people as possible. Project coordinators must consider applying only those with a proven track record, technical competence, and a deep understanding of the organization's needs. This way, they can foster a cohesive approach toward the system's development, encourage collaboration, and avoid the consequences of an uncoordinated approach.

Lessons Learned:

Align goals and establish a shared vision; It is essential to bring all stakeholders together to define and align the project's goals and objectives. A shared vision helps ensure everyone works towards a common purpose and minimizes conflicts fueled by personal agendas.

Create a cross-functional decision-making team;

Instead of including as many people as possible, establish a smaller, cross-functional team empowered to make critical decisions. This team should comprise individuals from different areas of expertise, representing various perspectives and capable of collaborating effectively.

Promote collaboration and effective communication; Encourage open dialogue and active listening among team members. Foster an environment where differing viewpoints are valued and constructive discussions occur. Effective communication channels and regular check-ins ensure everyone is on the same page.

What Should Have Been Done Differently:

To avoid the negative outcome caused by the decision to include too many people with strong personalities, the organization should have taken the following steps:

Establish a smaller, focused team; Create a decision-making team consisting of a balanced representation of stakeholders, including the steering committee, developers, and implementation team. This team should set the direction, make critical decisions, and ensure the project's goals are aligned.

Define a transparent decision-making process; Develop a structured decision-making process that encourages collaboration and prevents individual egos from dominating the discussions. This process should allow for the exploration of different ideas while ensuring decisions are made in the best interest of the project as a whole.

Facilitate team building and communication; Prioritize team building activities to help team members understand each other's perspectives, build trust, and develop effective working relationships. Promote open, transparent communication channels encouraging feedback, coordination, and information sharing.

By taking a more focused approach, encouraging collaboration, and establishing effective decision-making processes, the organization could have avoided the negative impact caused by selfishness and lack of coordination among team members. This would have increased the chances of delivering a more cohesive and successful system that genuinely addressed the organization's needs and challenges.

Expecting A Technologist/System Administrator To Understand Business Needs At the Level Required

Can introductory system administrators understand business requirements? **If your organization is like most, they can't.** It's not their fault. It's the business's fault; it's Finances fault, and it might be your fault if you go down this path.

Indeed, the role of Technology or System Administrators is rapidly changing in today's world, but this change is being forced on people that frequently don't have a choice and/or don't have those skills. So often, I interview CRM team leaders who state that their teams are being asked more and more to do tasks done only a few months ago by business people and/or administrative operations people. These people commonly shrug off this as "it's in the CRM, so it isn't our task/role anymore. This is becoming an epidemic and complete bullshit.

The Final Details

As a business, you may think having a CRM Administrator who can set strategy and understand business complexities is great. But don't count on it. It would be better to assign someone with the time, understanding, and experience to manage the CRM data and provide end-user support to address this issue.

An expert CRM Administrator should be able to take care of all the technical aspects of your CRM but don't expect them to understand the business strategy or be able to set it.

If you have a dedicated CRM administrator you rely on to serve as your CRM expert and make strategic decisions, they will not have time to do this. Make sure that you have clear expectations of the job description. If a CRM Administrator is expected to understand business needs and establish strategy, this will take away from their ability to build the system's foundation.

CRM software is a significant process and technical challenge for many users. A dedicated person for each business unit, team, and geography should be assigned to manage CRM data, create reports, customize the system, and provide end-user support to address this issue. Don't expect a CRM Administrator to be able to, or have the time, set strategy and understand business complexities.

11

Case Study: Growing System, Not Growing Team

Scene:

As a vivid snapshot that captures the essence of the dedicated professionals who tirelessly work to ensure the smooth operation of complex systems, Picture a dynamic office space teeming with system administrators, their desks adorned with technical books and certifications like badges of honor. As you step into the room, you're instantly enveloped in the symphony of sounds - the low hum of computers, the rhythmic clacking of keyboards, and the occasional exhalation of frustration when faced with a challenge.

Each administrator is engrossed in their work, channeling their expertise and experience towards delivering optimal performance from the intricate systems under their care. Their eyes dart across multiple screens, their fingers dancing across keyboards in a practiced rhythm. The air crackles with focused determination as they troubleshoot issues, meticulously scan logs, and fine-tune configurations to keep the systems humming.

These professionals are not just technologically proficient but also deeply understand the intricacies of the CRMs they support. Their shelves are adorned with technical books and certifications - tangible representations of their relentless pursuit of knowledge and their unyielding dedication to staying up to date with the latest advancements in the field.

Incident:

Over the years, I have witnessed a profound shift in the role of system administrators. As mega-systems such as CRMs and HR resource tools have risen, administrators confront new and unexpected challenges. What was once a clear-cut technical role has now expanded, encompassing many responsibilities. System administrators are now expected to act as business analysts, mini-project managers, system integration specialists for legacy systems, and even catalysts for business process overhauls.

This rapid growth in the scope of their duties has placed tremendous pressure and frustration upon system administrators. They are being stretched thin, burdened with the expectation to be masters of many disciplines without additional resources or support. While customers and users demand seamless operational support, they are left bewildered as to why the system lacks individuals capable of efficiently handling every aspect.

Conversely, business professionals are transitioning into technical roles as system administrators. While their business acumen can be an asset, they often lack the in-depth knowledge of system details and the logical mindset characteristic of a pure technologist. This deficiency of expertise invariably leads to inefficiencies and obstacles when attempting to troubleshoot and optimize the system.

Result:

The consequence of such a situation was a strained and overwhelmed workforce of system administrators. These dedicated professionals, who once thrived in their technical roles, struggled to keep up with the demands thrust upon them by their expanded responsibilities. As a result, efficiency suffered, frustration mounted, and cracks appeared in the once seamless support system.

Customers and users, unaware of the complexities and constraints administrators face, questioned why their requests weren't promptly and comprehensively fulfilled. The burden of their expectations weighed heavily on the already overwhelmed support staff, further exacerbating the situation.

However, the challenges did not end there. Administrators lacking technical expertise further compounded the difficulties in maintaining the system effectively. Certain intricacies were overlooked in the rush to adapt to their new roles, and optimal solutions were missed. The consequences of these oversights manifested in the form of suboptimal system performance, leaving both administrators and users grappling with the results.

In these instances, it became abundantly clear that unforeseen consequences can arise when the scope of responsibilities expands faster than resources and expertise can keep pace. The strain on the workforce of system administrators served as a reminder of the importance of providing adequate support, both in terms of resources and training, to handle the multifaceted demands placed on them.

As professionals, let us learn from these challenges and navigate our roles with caution and forethought. It is crucial to approach system administration with diligence, placing equal weight on technical expertise and the ability to effectively manage the expanded responsibilities of the territory. By doing so, we can mitigate the strain on administrators, ensure seamless support for customers and users, and maintain optimal system performance.

Let us be mindful of the potential pitfalls that can arise when roles expand beyond their initial boundaries. Through diligent planning, continuous learning, and a commitment to providing the necessary resources, we can empower our system administrators to thrive in their evolving roles and succeed in the ever-changing world of CRM.

Lessons Learned:

Recognize The Evolving Role Of System Administrators; Companies must acknowledge that the part of system administrators is no longer limited to technical tasks. They must appreciate administrators' expanded responsibilities, including business analysis, project management, and process optimization.

Provide Adequate Resources And Support; To ensure the success of system administrators, organizations should allocate additional resources and support. Hiring other staff, training in diverse areas, and fostering collaboration and knowledge-sharing can help administrators thrive in their expanded roles.

Balance Technical Expertise And Business Knowledge; Administrators transitioning from a business or non-technical background should receive focused training and support to develop the technical skills necessary for effective system administration. Balancing technical expertise and understanding business needs is crucial in optimizing system performance.

What Should Have Been Done Differently:

To address the challenges faced by system administrators and prevent the strain caused by their expanded roles, companies should take the following steps:

Provide Adequate Training And Development; Companies should invest in training programs tailored to the expanding responsibilities of system administrators. This training should focus on technical skills and business analysis, project management, and system integration. Such development opportunities empower administrators to excel in their multifaceted roles.

Build Cross-Functional Teams; Organizations should foster collaboration and knowledge-sharing among system administrators, business analysts, project managers, and other relevant stakeholders. Creating cross-functional teams encourages a balanced approach to problem-solving and enhances the effectiveness of system administration.

Regularly Assess And Adjust Staffing Levels; To address the concerns of customers and users regarding system support, companies should periodically evaluate staffing levels. Adequate staffing is crucial to ensure the availability of resources competent in handling various aspects of system administration and resolving issues promptly.

By recognizing the changing nature of the system administrator role, providing the necessary resources and support, and maintaining a balance between technical and business knowledge, companies can overcome the challenges and optimize system performance. It is essential to acknowledge the subtle differences in roles and skill sets and rely on something other than warm bodies in seats to address complex system needs effectively.

12

Wrong Team Structure And Talent For A CRM Effort

BUILDING THE RIGHT CORPORATE HIERARCHICAL STRUCTURE

Let's keep it straight. CRM is not a playground. You must play your cards right to get the most out of it. It's not about having the most expensive tools and solutions but having the proper structure and team to handle it. You can have a rocket without the right team, but it isn't getting off the ground!

Now, let's break it down, shall we?

The Hierarchical Blues

WHEN YOUR CRM development team doesn't have a transparent chain of command, problems pop up quicker than a hiccup. Lack of authority and clarity can cause serious stumbling blocks affecting deadlines, quality, and deliverability, translating all your potential CRM benefits into your worst nightmare.

Here's where it goes downhill:

Authority Crisis; Without clear-cut responsibilities, you are fast-tracked to delays and chaos. If your CRM Analyst acts like the CRM Manager and the Manager seeks to overstep, it's a soup of confusion. Everyone needs to know their place and tasks and keep their eyes on the prize.

Management Mishaps; If your team makes decisions on a whim, you're in for a bumpy ride. Being in management doesn't mean they know it all. Seeking advice is not a sign of weakness but an affirmation of being sensible.

Training Downfalls; Having a bunch of rookies in your CRM team can be like throwing a party but forgetting to invite guests: much effort with no payoff. You don't need to be a genius to know how bad that can get.

CRM-Specific Hierarchical Structure

THE STRUCTURE of your CRM team should be as systematic as the software itself.

Here's what you should aim for:

Role Clarity; This is the first rule of the CRM implementation club. You have specific roles like the CRM Manager, CRM developer, and CRM consultant, each with its power-punch of responsibilities.

Centralizing Decision Making; When it boils down to making a game-changing decision for your CRM strategy, it should rest with the senior CRM authority. No dilly-dallying here!

Flexibility in Structure; A rigid hierarchy can strangle your CRM operations. Maintain a semi-formal setting to allow the free flow of ideas. An idea can come from anywhere and should be valued, even by the intern.

Let's move on to the second act, shall we?

Building The Right CRM Team

Choosing the proper professionals for your CRM team is like setting up the critical pieces on a chessboard.

Here are a few pointers to consider:

Relevant Experience: Freshers may bring vigor to your team, but when handling CRM complexities, experience always wins.

Technical expertise: Want to skyrocket your CRM implementation? Bring people aboard who eat, breathe, and live CRM technicalities.

Interpersonal Skills: You need team members to bring harmony to your CRM strategic symphony. They can share, communicate and work as an integral part of the CRM melange.

The Divisional Liaison Connection

LASTLY, you cannot overlook the dotted lines. Divisional liaisons are your bridge to other divisions. They must have the authority and knowledge to represent their divisions and modify processes accordingly.

Divisional liaisons bring added advantages:

Implementing Changes; They can speed up implementing changes across divisions.

Swift Redressal; They can discuss, plan, and seek redressal for any CRM-related issues in their division.

Liaison Benefits; They assist in making sure the CRM activities are cohesive and aligned with the organization's goals across all domains.

CRM implementation is not a walk in the park. Establishing a solid hierarchy for training a robust team of professionals is about finding the right balance. Else, it's a titanic affair with no lifeboats in sight. But with smarts, a dash of humor, and an ocean of determination, it can be done and done.

SOMETIMES ASSIGNED **divisional liaisons are not a good fit** for the overall corporate advancement of CRM - sometimes, they either only know about their division and/or only care about their division. This may be acceptable, but more often than not, it is detrimental as you and your team will be constantly shifting gears to "deal" with them, and worse, the support functions will grow as a result. When these folks can't grasp the intricate conversation of CRM or the needs of their divisions, there's a gap that needs bridging.

Here's how to do it:

- Impart technical training or workshops to arm them with necessary CRM tactics.

- Have other better-aligned liaisons speak with them to help them see the big picture and/or teach them about their divisions.
- When insufficiencies persist, a conversation is vital. Address their incompetency head-on, but keep it pleasant and professional.
- When replacing a liaison isn't an option, organize regular meetings between the liaison, CRM team, and division heads to ensure a clear understanding of decisions and effective communication.

Switching divisional liaisons isn't a covert mission; it's a mature conversation.

Here's how to make it happen:

- Initiate the conversation with the divisional head, addressing the issues.
- If the divisional head looks the other way, escalate the matter to upper management. If done right, CRM implementation is a big win for the organization. Regardless of their station, you cannot let anyone be a roadblock for the entire firm.
- If you are forced to stick with the liaison, arrange for an intermediary who can facilitate communication and bring effectiveness to your CRM operations.

Addressing these concerns isn't pretty, but your CRM team can soar only through these seemingly tricky decisions. So gear up and unapologetically build a team that's made for CRM success.

Walking on Thin Ice: Upgrading or Removing CRM Staff

LET'S tackle the beast immediately- the dirty work of upgrading or removing your staff to meet the demands of your CRM development and support team. Nobody signed up to be the bad guy here, but business is business. Here comes the tricky part. When your staff doesn't match the CRM team's requirements, you have three avenues to explore:

Training; Nobody becomes a CRM wizard overnight. Hire a professional trainer, conduct workshops, and give them hands-on experience.

Transfer; If CRM is not their mojo, shift them to where they can excel and be assets. A square peg can't fit in a round hole; however hard you hammer it.

Termination; When all else fails, bite that bullet. But remember, a one-on-one honest conversation can save a lot of hard feelings.

NOTE: I have a ruthless-sounding practice affectionately called " *The Hudlow Hinky Rule*," which basically states that the second your deep gut tells you that a person isn't going to work, they aren't. I am sure this is not HR approved or recomended ! Sad but true that I have been running at a 100% track record for the past 20+ years. Consult with others and expedite a **fair path** for them to depart.

MESSING Up with the Staff Mix: Where the Trouble Brews

PLUGGING in the wrong team members into your CRM machinery is not unlike a bumpy ride on an unpaved road. Here's the rundown:

Needless delays; Lack of technical competence in CRM hinders workflow, churning out long, uncomfortable delays.

Collaboration goes haywire; The inexperienced find it tough to team up and get things done- It's not their cup of tea.

Quality dips south; When incompetence rules the roost, the functional quality of CRM output slips into the abyss.

The Life Map of Your CRM Staff

WHEN DEALING with retraining or termination of these staff members, here's a gold list of things to ponder upon:

Ease up on your training regimen; Not everyone's a nerd - When implementing a CRM system, finding the right balance between providing training and overwhelming your team is essential. While it's vital to ensure that everyone understands how to use CRM effectively, it's equally important to recognize that not everyone is a "nerd" or a tech-savvy expert.

Take a step back and consider your team's diverse skill sets and learning styles. Some team members may be more comfortable and quick to adapt to new technologies, while others may require more support and guidance. By easing up on your training regimen, you can create an environment that empowers everyone to succeed with the CRM.

Instead of bombarding your team with lengthy training sessions or complex technical jargon, focus on providing them with the essential information they need to get started. Break down the training into smaller, digestible modules that cater to different learning styles. Incorporate hands-on exercises, interactive activities, and real-life scenarios to make the learning more engaging and relatable.

Consider offering ongoing support and resources, such as step-by-step guides, video tutorials, and a dedicated help desk, to address any questions or challenges arising after the initial training. Encourage team members to share their knowledge and best practices, fostering a collaborative learning environment.

Remember, not everyone needs to become a CRM expert overnight. By striking a balance between training and understanding different learning needs, you can ensure that your team feels supported and equipped to utilize the CRM effectively in their roles.

Consider the cost-effectiveness of training versus

hiring new staff; When implementing a CRM, one of the critical decisions you'll face is whether to train your existing staff or bring in new hires. While new staff may seem convenient, weighing the cost-effectiveness of training versus hiring is essential. Training your current employees may be a more cost-effective and practical solution in the long run.

Training existing staff has some significant advantages. First, your current staff has invaluable knowledge of your company's objectives, vision, and systems. Hiring new staff to replace them might result in a steep learning curve, as they'll need time to get up to speed on internal processes and may struggle to integrate smoothly into the company's culture.

Secondly, training employees on a CRM can be much more cost-effective than hiring and integrating new staff. You can save time and money by utilizing your existing employee pool instead of bringing in new hires that require a significant investment in time and resources.

Furthermore, providing training and development opportunities to your current employees can increase their motivation and loyalty to the company, leading to higher productivity, job satisfaction, and employee retention rates. When considering whether to train or hire new staff, weighing the potential costs and benefits is essential. Training your employees on a CRM can be more practical, cost-effective, and beneficial when carried out correctly. By creating a comprehensive training plan and providing your staff with the necessary support and resources, you can ensure your team has the skills to fully leverage the CRM to drive growth and improve the customer experience.

Investigate if the employees genuinely want to shift gears to CRM (from another technology or field); When shifting gears to CRM, it's crucial to investigate whether your employees are genuinely interested in making the change. Implementing CRMs requires dedication and enthusiasm, so ensuring your team is on board before diving in headfirst is important.

Start by having open and honest conversations with your employees about the potential shift to CRM. Ask them about their thoughts, concerns, and interest in working with CRM technology. Encourage them to share their experiences with other technologies or fields and how CRM could enhance their work.

During these discussions, look for signs of genuine interest and excitement. Are they curious about CRM? Do they see its potential benefits to their roles and the company's overall success? Assessing their level of engagement is crucial as it will determine how smoothly the transition to CRM will go.

Remember that not everyone may be equally enthusiastic about the change, and that's okay. Creating a safe space where employees can voice their concerns and provide input. Address any apprehensions or doubts they may have and try to find ways to mitigate them.

Additionally, consider providing training and resources to support this transition. People are more likely to embrace new technology when they feel confident and equipped. Offer workshops, online courses, or access to CRM experts who can help your employees understand the ins and outs of CRM and how it can benefit them in their roles.

Remember, implementing CRM is not just about the technology itself. It's about people and their willingness to adapt and embrace change. By ensuring that your employees are genuinely interested in shifting gears to CRM, you'll set the stage for successful implementation and ultimately reap the benefits of improved customer relationships and business growth.

Ensure the termination doesn't hit your team's morale; Prepare them for it. - When making changes within your team or organization, such as terminating an employee, it's essential to approach the situation delicately and prepare your team for its impact on their morale. Sudden terminations can create anxiety and uncertainty among team members, ultimately affecting their motivation and productivity.

To mitigate the negative effects, being transparent and communicating openly with your team is crucial. Start by candidly conversing with the individuals directly affected by the termination. Explain the reasons behind the decision, being careful to maintain confidentiality when necessary. Emphasize that the termination is not a reflection of their own performance or abilities but rather a strategic decision made by the organization.

Consider the timing of the announcement as well. If possible, try to avoid making the termination announcement during particularly stressful or crucial periods of work. This can help minimize the immediate impact on the team's morale and allow them to adjust to the changes more smoothly.

Additionally, provide clear guidance and support during the transition period. Outline any changes to roles or responsibilities that may occur due to the termination and offer reassurance that the team will continue to be supported throughout the process. Encourage open dialogue and create an environment where team members feel comfortable expressing their concerns or asking questions.

Sometimes, providing opportunities for team members to voice their thoughts and concerns can help address any anxieties and build morale. Consider organizing team meetings or one-on-one conversations to allow individuals to express their emotions, ask for clarification, and discuss any challenges they may face.

Lastly, as a leader, it's crucial to lead by example and demonstrate resilience during challenging times. Show empathy and understanding, and encourage your team to focus on the future and the opportunities ahead. Remind them of the team's collective goals and the positive impact they can continue to make.

By taking these proactive steps to prepare your team for a termination and supporting them during the transition, you can help minimize the impact on morale and maintain a positive and productive work environment. Remember, change is a constant in any organization, and how you handle it can make a significant difference in the long-term success of your team and the overall morale.

The termination should be nothing less than dignified; no leaks, no humiliation, and absolutely no dirty laundry - When managing a team, tough decisions like terminating an employee can be a challenging but necessary part of the job. While it may be tempting to vent frustrations or share negative feedback with others, it's essential to approach the termination with the utmost professionalism, respect, and dignity.

First and foremost, it's critical to have a plan in place to ensure that the termination is carried out smoothly and discreetly. Limit the number of people with access to sensitive information about the termination, such as the reasons behind the decision, and ensure that all involved parties understand the importance of confidentiality. Any leaks or rumors can damage morale, poison the work environment, and undermine the organization's success.

Secondly, avoiding public humiliation or shaming of the employee in question is crucial. This includes not bad-mouthing the employee to others, whether within or outside the organization. Your reputation as a leader can be significantly impacted by any negative public perception of the termination, and it can be challenging to repair that damage. Instead, focus on the professional nature of the termination, honoring the employee's contributions and respecting their dignity.

13

Jack Of All Trades

In my years of experience as a professional in the field of CRM, I have often encountered the role of a "Jack of All Trades" to be pushed upon and/or expected of the system administrator. This is unfair and inconsiderate for them. These individuals take a hands-on approach to tackling the most common CRM security, privacy, integration, and performance issues. A single administrator oversees distinct responsibilities such as programming, coordinating, and managing. Wow!

Undoubtedly, the role of a CRM system administrator is crucial for the successful customization of the CRM application to meet specific business requirements. However, it is critical to recognize that achieving success in this realm often requires the help and guidance of the business units themselves. Without collaboration and input from these units, the administrator's efforts may fall short.

A system administrator must comprehensively understand CRM software and sales, brand management, and business operations to fulfill their responsibilities effectively. The breadth of knowledge required for this role can be extensive and demanding. Expecting one individual to possess such a diverse skill set and expertise is a tall order.

It is crucial to set realistic expectations for a system administrator. Before delving into this role, it is essential to ask oneself whether the emphasis of the administrator's job should lean towards technical proficiency or a deep comprehension of business processes. Striking a balance between the two is vital. It would be best if you were prepared to determine the split, whether 60/40, 70/30, or another distribution that suits your unique needs.

Furthermore, it is crucial to remember that system administrators may lean towards a strong focus on technology. This specialization may arise due to their extensive training and experience, making them more comfortable and effective in handling technical aspects. Awareness of this potential bias is crucial to ensure you have the right expectations and are prepared to address any gaps or limitations.

While leveraging a "Jack of All Trades" system administrator can be a valuable asset to your CRM implementation, it is vital to approach this role with pragmatism and caution. Recognize the need for collaboration with business units, set realistic expectations regarding the position's technical and business process aspects, and acknowledge that specialization in technology may be the natural inclination of the system administrator. By understanding these dynamics, you can optimize your CRM implementation journey and set your organization up for success.

14

Case Study: Gimme A Promotion

I HAVE BEEN HERE FOR FIVE WHOLE MONTHS

Scene:

Picture yourself stepping into the office space of one such company, a haven of collaboration and innovation. The atmosphere is open and airy, with sunlight streaming through the windows, casting a warm glow on the bustling activity. The office buzzes with the hum of creativity as teams of technologists forge ahead to stay ahead of the ever-growing demands of users and customers.

The company recognizes the need to bolster its technical capabilities in this dynamic setting. They are aware that their success hinges on their ability to implement and support new systems, catering to the changing needs of their users and customers. To fuel this pursuit, the decision has been made to hire new technologists with the expertise to navigate the complex terrain of cutting-edge technologies.

The company is driven by anticipation and eagerness, knowing that the right technologists will play a pivotal role in shaping their future. They seek individuals who can seamlessly integrate these new systems, maximizing efficiency and productivity. These technologists must possess technical prowess and a deep understanding of user and customer needs.

As you immerse yourself in this scene, it becomes evident that this company recognizes the importance of staying ahead in the ever-advancing world of technology. They understand that hiring the right talent is crucial to their success, enabling them to meet the growing demands of their users and customers. By proactively seeking out talented technologists, they demonstrate their commitment to evolving with the ever-changing industry landscape.

Incident:

In the current landscape, where CRMs have become a sought-after commodity, there has been a notable surge in new technologists eager to embark on their professional journeys. Fresh out of school and hungry for job opportunities, they are drawn to the allure of being part of the CRM action. However, a troubling trend has emerged among these aspiring technologists that warrant attention and concern.

To my disappointment, it has become apparent that many of these newcomers arrive expecting to achieve rapid advancement in title and remuneration. Some anticipate ascending the ranks and reaping the rewards even after a **few months** on the job. The reasons behind this mindset remain a perplexing mystery to me. Yet, there is no denying the impact it has on the industry.

Unfortunately, this seemingly promising ambition poses substantial challenges for companies. Faced with an influx of employees with possibly unrealistic expectations of swift progress, organizations grapple with the consequences. When these new hires discover that advancement is not as immediate as they envisioned, many abandon ship prematurely.

This perpetuates a cycle of short-term employees who contribute little to the company before moving on. Such rapid turnover drains financial resources and places an unnecessary burden on trainers who invest time and effort in grooming these individuals. It is a disheartening situation that hampers productivity and hinders the potential growth of individuals and the organizations they join.

As seasoned professionals in the realm of CRM, we must recognize the risks associated with this trend and remain vigilant against its adverse effects. It is imperative for hiring managers and executives to set realistic expectations during the recruitment process, ensuring that prospective employees understand the opportunities for growth and advancement within the organization. By fostering transparency and open communication about career progression, companies can manage the aspirations of new technologists more effectively, minimizing the likelihood of unrealistic demands.

Result:

The repercussions of the prevailing trend of technologists expecting rapid promotion were far-reaching, causing detrimental outcomes for numerous companies. High turnover rates became the norm as these ambitious individuals quickly grew disillusioned with the reality of their career progression. This constant influx and exodus of employees profoundly impacted productivity, which plummeted to alarming lows.

The financial implications were equally significant, with companies bearing substantial losses. The resources invested in recruiting, training, and onboarding these technologists were squandered as they departed after merely a few months. This revolving door of talent drained financial reserves, eroded team morale and hindered the development of cohesive work environments.

Perhaps the most detrimental consequence was the loss of valuable institutional knowledge. With an ever-changing workforce, companies struggled to retain the expertise and industry-specific insights acquired over time. As experienced employees walked out the door, their institutional knowledge went with them, leaving a void that couldn't quickly be filled. This lack of continuity disrupted workflows and diminished the collective wisdom that only comes with years of hands-on experience.

In some cases, the toxic nature of this revolving door scenario further exacerbated the problems. The constant turnover cultivated an environment of instability and uncertainty within the organization. Morale among the remaining employees suffered, and disillusionment permeated the workplace.

This costly aftermath serves as a sobering reminder of the importance of proactive measures to address this trend. It highlights the critical need for companies to cultivate an environment where realistic expectations and transparent career paths are established from the outset. By nurturing a culture that values growth and development over immediate promotions, businesses can mitigate the detrimental effects of high turnover rates, low productivity, and financial losses.

The consequences resulting from the trend of technologists expecting rapid promotion were far-reaching and severe. The detrimental impact on turnover rates, productivity, financial stability, and institutional knowledge cannot be understated. It behooves organizations to learn from these experiences and implement strategies prioritizing long-term career growth, fostering stability and success for the individual technologists and the companies they serve.

Lessons Learned:

Set Clear Expectations; From the outset, companies should establish clear expectations with new hires regarding job responsibilities, promotion pathways, and the timeline for reaching these objectives. Effective communication can help avoid misunderstandings and ensure that both parties are on the same page.

Analyze Motivations Before Hiring; To avoid high turnover rates and wasted resources, it is vital to analyze the motivations of potential hires before making an offer. Companies should use sophisticated tests and questionnaires to identify personality traits and work styles to help them thrive in the company culture.

Create A Robust Training And Development Program;

Companies can prevent employees from leaving prematurely by investing in their growth. Providing relevant training and development opportunities can help keep employees engaged throughout their tenure.

What Should Have Been Done Differently:

To avoid the negative consequences of the trend of technologists expecting quick promotions and leaving without contributing much, companies must take steps to address the issue. The following steps could help improve the situation:

Implement a fair and transparent promotion system; Companies should establish a clear and fair promotion system that communicates the criteria for advancement. This system can help manage employee expectations and encourage long-term commitment.

Conduct thorough job interviews; Employers should conduct comprehensive, critical behavior interviews with job candidates to identify their goals and motivations. This can help avoid problematic hires and identify better long-term fits for the company.

Provide mentorship and leadership training; Companies should consider investing in leadership training and mentorship programs to help employees learn how to address difficult situations and develop a growth mindset. These programs can foster a culture of long-term commitment and learning.

Require maturity, experience, aptitude, ability and realistic thinking in the people you hire 😐 AND keep in check what you will require of them 😳.

While the trend of technologists expecting rapid promotion may sound promising initially, it can create significant problems for companies. To create an energetic and productive work environment, it is essential to establish clear expectations, analyze motivations before hiring, and invest in training and development programs.

15

Migration From One Behemoth To Another

Ah, the infamous concern of migrating from a large CRM to their competition in the future, only to face roadblocks and resistance. It's a legitimate worry that many have grappled with. Let me shed some light on this for you.

First, let's acknowledge the possibility of such a situation. While it's not unheard of for a large CRM vendor to make it difficult for customers to transition to a competitor, it's rare. Most reputable CRM providers understand that customer satisfaction is vital to their success and maintaining a positive reputation in the market. They know that locking customers in or making it arduous to migrate can damage their brand and relationships long-term.

However, there have been cases where vendors have imposed obstacles when companies try to switch CRM platforms. These tactics could range from making it challenging to export data in a compatible format to imposing hefty fees or penalties for terminating contracts prematurely. It's a frustrating scenario, no doubt.

So, what can you do to mitigate this risk? You can take a few steps to protect yourself and ensure a smoother transition if you switch CRM platforms down the road:

. . .

DO YOUR DUE DILIGENCE; Research their track record and reputation before committing to a CRM provider. Look for customer testimonials, reviews, and ratings to understand their customer-centricity. It might be a red flag if you find a pattern of customers facing difficulties in migrating.

Review The Contract; Carefully read through the terms and conditions of the CRM agreement. Pay attention to any data ownership, termination, and migration clauses. Look for any hidden fees or penalties that may apply if you decide to switch vendors.

Negotiate Contract Terms; If you have concerns about future migration, discuss them upfront with the vendor during contract negotiations. Seek assurances that they will facilitate a smooth transition if you decide to switch providers. Get these assurances documented in the contract to protect yourself.

Backup Your Data Regularly; One way to safeguard against a CRM vendor making it difficult to export your data is by periodically backing up your CRM data independently. By having your data backed up and accessible, you have more control and flexibility in case you need to migrate quickly.

Stay Informed And Plan Ahead; Keep an eye on industry trends and the CRM market landscape. Stay knowledgeable about potential changes or shifts in your CRM provider's policies that might impact data migration. By being proactive and planning ahead, you can preemptively address any challenges that may arise.

WHILE THERE ARE some valid stories about vendors making it difficult to migrate, it's not an everyday occurrence. By doing your homework, negotiating the correct terms, and taking proactive measures, you can minimize the risk and ensure a smoother transition if needed. Don't let the fear of migration hold you back from choosing the CRM solution that best fits your needs and business goals.

Let me set the record straight: Mega-CRM vendors are drooling at the opportunity to woo their competitors' customers. Picture a pack of wolves salivating over the sight of a lost sheep - it's intense.

Their opening move? Throw themselves at you with all the panache of a Hollywood stunt double, coupled with deep discounts on transition services. My point is simple. These mega CRMs got to be mega because they are good and connect themselves to other things. Transition is never fun, but we will all have to do it eventually. Would you instead do a 1:1 migration or a many: many?

16

Case Study: Mature Operations And System Operations/New Responsibilities

Scene:

Imagine a bustling division that makes a good profit for the larger firm, bustling with activity. This division's marketing team's role is to devise creative campaigns that showcase the company's products and services to target audiences. Ideally, these marketing departments should work fluidly with CRMs to operationalize their campaigns and gain insights into their audiences. However, from time to time, this marketing team can become frustrated with the operational and technical limitations of CRMs. One such incident is detailed below.

Incident:

A divisional marketing had grown tired of the rigid process flow into campaigns within their CRM platform. They saw process flows as restrictive and limiting their creativity. So, in defiance of the established processes, Marketing decided to bypass the system entirely and use their own add-ons, bolt-ons, and offline tools that better aligned with their campaign goals. This approach was a flagrant violation of company policies, but some within marketing saw it as necessary to achieve their objectives.

Result:

The results of the Marketing team's approach were problematic on several levels:

Data Integrity Issues; By bypassing the CRM process flow, crucial data was lost, and critical metrics were not accurately tracked.

Decreased Operational Efficiencies; By ignoring established processes, Marketing created inefficiencies and increased potential errors in their workflows.

Security Risks; Using offline tools and add-ons that had not been vetted by the IT team created security risks.

Lessons Learned:

This incident revealed several critical lessons about CRM and marketing operations:

Clear Communication; Open communication between marketing and operational teams is essential to ensure campaign creativity while maintaining data security and operational efficiency.

Avoid Shadow Teams; IT, Marketing, or any other department, should not use their own off-line tools and add-ons without first seeking IT's approval.

Established Processes Are There For A Reason; The processes established within CRMs are there for a reason, so ignoring or bypassing them only creates more problems.

What Should Have Been Done Differently?:

Marketing could have achieved its goals within the CRM process flow by:

Seeking Collaboration; By seeking collaboration with operational teams, Marketing could have found a way to work within established processes while maintaining creative freedom.

Following IT Policies; Marketing, or any other department, should have followed company policies and protocols regarding offline tools and add-ons.

Customizing CRM Processes; By working with IT, Marketing could have customized the CRM flows to better align with their campaign goals while maintaining data integrity and security.

Proper workflow and process ownership are crucial to establishing an operational flow that works for your business, not against it.

17

More CRM Technology Is Better

Believing That Having More CRM Technology Is Better

There's a common fatal misconception held by many executives regarding the success of CRM solutions. They tend to believe it all comes down to having the latest, cutting-edge technology and advanced infrastructure. But there's more to the story than just fancy tech.

Understanding the importance of linking process improvement to technology is crucial, especially in CRM solutions. Technology does play a pivotal role in managing customer data and streamlining processes. It provides us with the tools and capabilities to store, analyze, and access information more efficiently. But here's the thing – technology alone is not enough to guarantee success. **You will fail if processes are not improved before or in conjunction with the implementation of any CRM.** As I have previously mentioned, you may already have (and likely do) have a CRM "system", it just does not meet your needs or is so disconnected that it has no resemblance of a CRM 😎. My point being that if you did everything in this book but didn't replace the technology, then you still would have moved your CRM situation significantly down the path of progress.

By aligning process improvement with technology, organizations can unlock the full potential of their CRM objectives and drive sustainable growth.

Think about it this way; technology is just a means to an end. It's the vehicle that enables us to achieve our CRM goals. But if we neglect the underlying business processes that support and leverage that technology, we're missing out on a whole world of possibilities.

It is unfortunately pervasive that companies, after investing heavily in the latest CRM technology, expecting it to revolutionize their sales and customer service processes, too quickly realize that the technology alone didn't magically solve their problems. The real breakthrough came only after much frustration and angry words were shouted across a boardroom table when they took a step back and looked closely at their existing processes. They identified inefficiencies and bottlenecks, and through some back-peddling and system redesign, a systematic approach to process improvement, they aligned their business processes with the capabilities of the CRM technology. That's when the true transformation occurred!

When implementing a CRM solution, **never focus on the technology itself.** Take the time to evaluate your existing processes, identify areas for improvement, and align them with the capabilities of the technology. By doing so, you can ensure that every piece of the puzzle is in its proper place, working together harmoniously to achieve your CRM objectives.

NOTE: Success in CRM isn't just about having the flashiest tech or the most advanced infrastructure. It's about striking that perfect balance between technology and business process improvement. When you align these two aspects, you set yourself up for effective CRM implementation, enhanced customer experiences, and sustainable growth.

Some Risks of Not Linking Process Improvement to Technology:

INEFFICIENT WORKFLOW; Without integrating process improvement into technology implementation, organizations may risk running inefficient workflows. While technology can automate and streamline specific tasks, if the underlying processes are not optimized, the full potential of the CRM may not be realized. Inefficiencies and bottlenecks can hinder productivity, lead to errors, and cause employee frustration.

Data Silos And Inconsistencies; Neglecting to align process improvement with technology can result in data silos and inconsistencies. When different teams or departments adopt separate processes not integrated into the CRM, data may not flow seamlessly across the organization. This can lead to duplication, inaccurate reporting, and challenges in gaining a holistic view of customer interactions.

Missed Opportunities for Standardization; Implementing a CRM provides an opportunity to standardize processes across the organization. However, failing to link process improvement efforts to technology can result in missed opportunities to create standardized workflows, guidelines, and best practices. This leads to inconsistency in how customer data is captured, interpreted, and utilized, making it difficult to achieve a unified customer experience.

Limited Scalability and Adaptability; Organizations may struggle to scale their CRM capabilities and adapt to changing business requirements without focusing on process improvement alongside technology. As the organization grows, processes must be flexible and adaptable to accommodate evolving customer needs and market trends. Disconnecting process improvement from technology hampers the ability to make agile adjustments and limits the system's long-term viability.

Resistance to Change and Low User Adoption; People are creatures of habit. Employees may resist the change when technology is introduced without considering process improvement. They may find it challenging to navigate the new system or struggle with altered workflows. This resistance can lead to low user adoption rates, rendering the technology investment less effective and inhibiting the organization's ability to derive maximum value from the CRM solution.

Linking process improvement to technology: organizations can mitigate these risks and realize the rewards of a well-aligned CRM strategy such as:

- **Streamlined and efficient workflows** that maximize the potential of the CRM
- **Consistent and accurate data** across teams, enabling better decision-making.
- **Standardized processes** that enhance the customer experience.
- **Scalability and adaptability** to accommodate business growth and market changes.
- **Higher user adoption** rates and increased employee engagement.

NOTE: **Technology alone is never the complete CRM solution:** By combining process improvement with technology, organizations can unlock the true power of CRM solutions and achieve their objectives more effectively. Embrace the symbiotic relationship between process improvement and technology, and let them fuel your organization's success. The CRM journey will have streamlined processes, optimized workflows, and satisfied customers.

IN A WORLD that increasingly revolves around technology, it is easy to assume that CRM solutions must rely heavily on state-of-the-art technological advancements. However, this is not always the case. Effective CRM encompasses various aspects of operation, and achieving CRM goals can be realized without significant infrastructural transformations. This section will focus on the role of business process improvement in achieving CRM goals. It will also highlight some challenges when implementing CRM solutions and how they can be overcome.

18

Business Process Improvement

The imperative of business process improvement is often overlooked. Let's dive deeper into the importance of business process improvement and the challenges that come with it. Improving existing business processes can often be sensitive, especially when dealing with long-standing processes. It's akin to approaching a proud business leader and suggesting that their baby is not the cutest thing in the world. As such, it's crucial to approach these situations delicately and strategically.

One common challenge with process improvement is overcoming the emotional attachment that business leaders have with their processes. After all, these processes were often created by the leaders themselves. It can be challenging to convince a business leader that the process they designed needs improvement or even replacement. Thus, it's essential to approach the conversation with empathy and to listen carefully to the leader's concerns and insights.

Another challenge is identifying the most critical processes to improve. In most organizations, countless processes and systems are in place, some of which may have been around for years. Not all processes are created equal, and it's important to prioritize which ones to improve based on their impact on the organization's overall goals, efficiency, and customer experience. An experienced mentor will help you recognize the processes that should be prioritized for improvement based on the business areas that need attention.

Yet another challenge is getting everyone on board with the process improvement effort. It's vital to ensure that all stakeholders are engaged in the process improvement journey and understand the "why" behind it. Thorough communication and employee involvement will help to foster buy-in and encourage participation from all parties.

While the imperative of business process improvement is often overlooked, it's critical to achieving your organization's goals. By approaching the task with empathy, prioritizing the most impactful processes, and getting everyone involved, you'll set yourself up for success. Remember, it can be challenging to do this work, and you may encounter some hurdles along the way. However, you can overcome these obstacles and create real change with a supportive mentor and the right mindset.

All of this being considered is essential because process improvement is generally not as glamorous as technological advancements. Still, it can be just as effective in improving CRM processes and achieving desired results. Business process improvement encompasses all the activities required to complete your CRM goals, including:

CRM solutions without the heavy reliance on technology can be ingrained in the organizational culture focused on meeting the specific needs of individual customers.

Organizations can foster an environment that supports efficient CRM practices without investing heavily in technological infrastructure by encouraging and empowering employees to become more conscious of customer requirements, wants, and preferences.

Employees can adopt customer-centric mindsets through proper training, guidance, and support. As a result, businesses can deliver tailored services and products that align closely with customers' expectations.

Decision-makers must acknowledge and embrace that focusing solely on technological solutions or assuming that high-tech implementations are superior to low-tech alternatives can prove expensive and ineffective.

19

Over-Customization

There are dangers if you over-customizing CRMs. While a customized CRM solution can be a good thing, too much customization can do more harm than good. Here are a few things to remember:

Over-Customization; can lead to unnecessary complexity, making it harder to streamline operations and simplify tasks. Instead of automating and streamlining processes, an over-customized CRM solution can become cumbersome, requiring more resources to keep operational. This can include specialized personnel such as data scientists, analysts, and high-priced contractors to clean up the effects of years of unnecessary complications.

Complexity And Streamlining; Over-customizing a CRM can lead to unnecessary complexity, which makes it harder to streamline operations and simplify tasks. When considering customization, it's crucial to evaluate whether it improves efficiency or adds unnecessary layers of complexity.

Resource Allocation; An over-customized CRM solution may require additional resources to keep it operational. This can include specialized personnel, such as data scientists and analysts, who must manage the complex architecture resulting from over-customization. It may also involve hiring high-priced contractors to clean up the effects of years of unnecessary complications. Before diving into customization, consider the long-term resource implications and whether it aligns with your organization's budget and goals.

Automation And Efficiency; A CRM aims to automate processes and streamline operations. However, over-customization can hinder automation efforts and make tasks more labor-intensive. When customizing your CRM, carefully consider whether the changes align to automate processes and improve efficiency.

Long-Term Maintenance; Over-customization can make a CRM more challenging to maintain and support over time. Consider the long-term maintenance costs and resources required to keep the customized system running. It's essential to strike a balance between customization and ease of maintenance.

Requirements vs. Preferences; When customizing a CRM, it's important to differentiate between genuine business requirements and individual preferences. Evaluate whether the proposed customizations genuinely address critical business needs rather than accommodating personal biases that may lead to unnecessary complexity.

DEEP BACK-END CONFIGURATION can cause more trouble than it's worth. When administrators tailor functions to their roles or tasks, it can create an intricate IT landscape prone to complexities as the organization grows. Even worse, when multiple iterations of CRMs are acquired, each catering to specific purposes and functions, this can result in data silos that make it difficult for different departments to keep track of customer information.

IT Landscape; Deep back-end configuration can lead to an intricate IT landscape that is difficult to navigate and manage. When administrators tailor functions to their roles or tasks, it can lead to a complex and customized environment that is difficult to maintain.

Scalability; As the organization grows, an intricate IT landscape can become even more challenging. When considering customization, evaluating whether it is flexible and scalable enough to adapt as the organization's needs evolve over time is crucial.

Multiple Iterations; When multiple iterations of CRMs are acquired, each catering to specific purposes and functions, this can result in data silos that make it difficult for different departments to maintain a unified view of customer information. This hinders effective communication, collaboration, and decision-making among various teams and can result in wasted financial resources.

Data Management; A customized IT landscape can result in more data to manage, which can be overwhelming if poorly organized. When customizing your CRM, it's essential to consider how data will be organized, managed, and accessed across different departments.

Data Synchronization; When multiple iterations of CRMs are used across different departments, data synchronization can become challenging. Ensuring that all data is synchronized and up-to-date across all systems is essential for maintaining a complete view of the customer journey.

SOMETIMES LESS REALLY IS MORE; It's essential to take a balanced and measured approach to CRM customization, combining the strengths of technology with robust business process improvements. By ingraining customer-centric attitudes within the organizational culture, actively seeking ways to streamline operations, and carefully considering the appropriate level of technological implementation, organizations can create CRM solutions that are both efficient and cost-effective.

Technology And Business Process Improvements; CRM customization should not focus solely on technology but also on improving business processes. It's important to assess existing processes, identify bottlenecks, and seek ways to streamline operations before implementing customization. By aligning technology with business process improvements to create a more efficient CRM solution.

Customer-Centric Attitudes; Ingraining customer-centric philosophies within the organizational culture is fundamental for CRM success. A deep understanding of customer needs and ways of operating should drive customization. By placing the customer at the center of CRM customization efforts, organizations can ensure that the customization enhances the customer experience and adds value.

Streamlining Operations; Customizing CRM should involve thoroughly evaluating existing processes to identify areas that can be simplified. Organizations can maximize the benefits of customization and improve overall productivity by eliminating redundancies, automating manual tasks, and creating efficient workflows.

Appropriate Technological Implementation; It's crucial to carefully consider the level of technical performance required for CRM customization. Sometimes, less is more. Selecting only the necessary features and functionalities that align with business needs and goals can lead to a more cost-effective and manageable CRM solution. It's essential to avoid overloading the system with unnecessary complexities that can hinder adoption and usability.

Efficiency And Cost-Effectiveness; A balanced and measured CRM customization approach can help create efficient and cost-effective solutions. Organizations can reduce costs, improve productivity, and increase customer satisfaction by optimizing processes and purposefully integrating technology.

CRM IMPLEMENTATION SHOULD NOT BE VIEWED as a one-off project but rather as an ongoing process. Organizations must continually evolve their customer experience strategies to stay relevant in today's marketplace.

Continuous Improvement; Organizations must continuously evaluate their CRM strategies and be open to making necessary improvements. This ensures that the customer experience remains top-notch and that the organization continues to meet customer needs and expectations.

Keep pace With Technology; Customer expectations are constantly evolving, as are the technologies used to meet those expectations. Organizations must stay up to date with these technological advancements to effectively implement CRM strategies that best meet the needs of their customers.

Data-Driven Insights: Organizations must continuously leverage data insights to improve their CRM strategy. This involves a deep understanding of customer behavior, needs, and preferences, gleaned from customer data analysis in order to make informed decisions.

Collaborative Approach: A successful CRM strategy involves cross-functional collaboration across all departments within the organization. When all departments align and work in the same direction, they can better serve customers, anticipate their needs, and deliver exceptional service.

Flexibility And Adaptability: The CRM strategy must reflect these changes as customer needs evolve. Organizations must remain flexible and adaptable to remain competitive and best meet the needs of their customers.

THE ALLURE of CRM customization can be quite enticing! I've been there and understand the temptation to go all-in, making every little tweak and modification to align your CRM with your organization's unique needs perfectly. But let me share some hard-earned wisdom with you: be cautious of the pitfalls of over-customization.

Finding the right balance is vital. It's not about abandoning customization altogether but following a balanced approach combining technology with business process improvements. This harmonious blend creates CRM solutions that drive growth and foster a positive customer experience.

Technology is a powerful tool in the CRM universe, offering various customization options. It's easy to get lost in the vortex of customization, tweaking every little detail until the CRM resembles a masterpiece crafted just for your organization. However, it's essential to approach customization with a discerning eye.

Remember, technology is a means to an end, not the end itself. Evaluate whether each customization aligns with your organization's goals and objectives. Ask yourself if they truly serve a purpose or are merely cosmetic changes. Sometimes, simplicity is the key to success. But it doesn't stop there. Business processes are the backbone of any CRM. Before diving headfirst into customization, take a step back and evaluate your existing processes. Are there inefficiencies that need to be addressed? Bottlenecks that need to be cleared? Manual tasks that can be automated? Improving your underlying processes sets the stage for effective CRM customization. Don't embark on this journey alone. Involving your stakeholders is crucial. Seek input from key team members, department heads, and your customers. Their perspectives will provide valuable insights to ensure that customization decisions are well-informed and align with the needs of the entire organization.

As your organization grows, you'll need a CRM that can keep pace with that growth. Scalability is vital. Customization efforts should be future-proof, capable of adapting to changing needs without tying you down in a tangled mess. Stay agile and plan for the long term.

And let's not forget about training and support. Customizing your CRM introduces new features and functionalities your team needs to master. Make sure you provide adequate training and ongoing support to ensure smooth adoption. A custom CRM is only as powerful as those who use it.

Finding the right balance is crucial. It's not about discarding customization entirely but approaching it with wisdom and caution. Blend technology with business process improvements, involve your stakeholders, plan for scalability, and provide training and support.

20

The Impact Of Data Privacy And Security On CRM Success

In the modern business world, data-driven strategies have taken center stage, making robust CRM systems more essential than ever. These complex platforms are instrumental in harnessing a wealth of customer interactions and company information, leveraging them for strategic competitive advantages. However, a parallel obligation for rigorous data privacy and security prevails alongside such immense value. These two factors, therefore, emerge as critical ingredients in the successful adoption and efficacious ongoing use of CRMs.

Primarily, the information ensconced within CRM solutions typically spans sensitive details about customers and proprietary company data. On the one hand, from a client's standpoint, these details encompass contact information, behavioral dispositions, purchasing patterns and propensities, personal preferences, and interactions with the brand. On the other hand, regarding company data, these systems streamline various confidential aspects, ranging from high-level strategic decisions to operational protocols. Given its substantial breadth and depth, the data in these systems is not just valuable - it's sensitive and could be potentially damaging in the wrong hands.

Therefore, ensuring that such wide-ranging and impactful data is safeguarded and responsibly managed is not just a superficial adherence to legal and ethical mandates. Instead, it demonstrates an earnest commitment to upholding an organization's obligations towards its stakeholders - an integral component of solidifying trust among employees and customers. Hence, data privacy and security are the fundamental pillars supporting any CRM's long-term reliability and success.

Diving deeper into the significance of these two aspects, one must primarily understand the importance of privacy in the context of a CRM. Despite residing under the security umbrella, data privacy has a unique individual standing in its relationship with CRM solutions. It refers to empowering clients with the right to control their personal information, dictating how it's gathered, stored, and used. In the era of rising protective regulations like the General Data Protection Regulation (GDPR) or the California Consumer Privacy Act (CCPA), businesses are tasked with more than ever the onus to uphold these rights.

Privacy concerns include complying with these stipulations to dodge hefty legal penalties. They also play a massive role in nurturing customer relationships. Clients today are increasingly aware and vocal about their data privacy rights, and any perceived breaches can lead to potentially irreparable tarnishing of the company-client relationship. Hence, demonstrating an active commitment to data privacy can elevate customer trust and loyalty - a treasure for any client-focused business.

On the flip side, I saw the significant impact of poor data privacy practices past several headline-making mishaps. These incidents have undoubtedly underscored the financial and reputational losses businesses can face due to mishandled or inadequately protected customer data. Therefore, demonstrating an ongoing commitment to data privacy is a non-negotiable aspect of modern business operations.

Parallel to privacy is the essentiality of robust data security measures. Given the sensitive and valuable nature of the data housed by CRMs, these platforms can become tempting targets for malicious cyber threats. The potential fallout of a data breach is devastating and far-reaching. It encompasses financial losses and legal penalties, significantly damaging an organization's reputation.

Implementing stringent data security protocols and measures ensures this extensive, valuable data repository is well protected. Security doesn't just prevent unauthorized access or potential breaches. An effective security solution will also foster a more confident and trusting relationship with the users of the CRM – be it employees accessing it for daily work or customers whose data is within the system. By ensuring that their sensitive data is well-shielded against potential vulnerabilities or threats, you foster an environment of trust that enables smoother CRM adoption and confident ongoing use.

Yet, robust data privacy and security protocols alone do not guarantee seamless CRM adoption. An equally important part lies in openly communicating these measures to the concerned parties, addressing any potential apprehensions they might harbor. For instance, many employees are commonly concerned about the perceived risk of confidential information leaks via CRM.

To reduce such fears, businesses must ensure transparency in their data privacy and security policies and proactively communicate these to their employees. Regular training and open-floor discussions can cultivate a mindset wherein employees view the CRM as a trusted ally rather than a potential security threat. Also, a well-articulated and enforced data governance policy can provide clear guidelines on data access, handling, and sharing, offering additional reassurance to the employees.

Meanwhile, on the customer front, businesses can leverage various channels like email newsletters, social media updates, or even personalized letters to communicate their commitment to data privacy and security. Offering clients easy-to-understand explanations about how their data is being used and protected secures their consent and empowers them with knowledge, instilling a heightened sense of trust.

The role of data privacy and security in the adoption and

ongoing use of CRMs is immense and enduring. These factors stand center stage in successfully integrating these data-rich systems into everyday business operations. By employing robust data privacy and security protocols and ensuring open, transparent communication about these structures, businesses can foster a climate of trust around CRM solutions that encourages their successful adoption and optimal utilization.

The Role of Data Privacy and Security in CRM Adoption and Ongoing Use

Data privacy and security play a large role in the adoption and ongoing use of CRMs for several reasons:

Building Trust With Employees; Employees are more likely to adopt and use a CRM if they trust that it will protect their personal information and that of their customers. Ensuring the CRM has robust privacy and security measures can help build this trust and encourage employees to use the system confidently.

Building Trust With Customers; Customers are increasingly concerned about the privacy and security of their data, and businesses must demonstrate that they take these concerns seriously. A CRM with strong privacy and security measures can help to build trust with customers, as they can be confident that their data is being handled responsibly.

Compliance With Data Protection Regulations; Businesses must comply with data protection regulations, such as the General Data Protection Regulation (GDPR) in the European Union or the California Consumer Privacy Act (CCPA) in the United States. Ensuring that the CRM adheres to these regulations is crucial to avoid fines and legal penalties and maintain a positive reputation.

Preventing Data Breaches And Cyberattacks; Data breaches and cyberattacks can have severe consequences for businesses, including financial losses, reputational damage, and loss of unilateral trust. A CRM with robust privacy and security measures can help to prevent these incidents by ensuring sensitive data is safe and reducing the risk of unauthorized access.

Ensuring Data Accuracy And Integrity; Data privacy and security measures can help to ensure that the data stored in the CRM is accurate and reliable. By protecting data from unauthorized access and manipulation, businesses can be confident that the insights and analytics generated by the CRM are based on high-quality data.

Addressing Data Privacy and Security Concerns in CRMs

BUSINESSES CAN TAKE several steps to address potential data privacy and security concerns and ensure the success of the CRM, including:

Selecting A CRM With Robust Privacy And Security Features; When choosing a CRM, businesses should prioritize privacy and security features, such as data encryption, access controls, and regular security updates. These features can help to protect private data and minimize the risk of data breaches and cyberattacks.

Implementing Strong Access Controls; Limiting access to sensitive data within the CRM to only those employees who need it for their job can help to prevent unauthorized access to data breaches. Businesses should implement strong access controls, such as role-based access and multi-factor authentication, to ensure that only authorized individuals can access sensitive information.

Training All Parties On Data Privacy And Security Best Practices; Employees play a critical role in maintaining data privacy and security. Regular training on best practices, such as strong password management and recognizing phishing attempts, can help reduce the risk of human error.

Developing And Enforcing Data Privacy And Security Policies; Businesses should develop clear and comprehensive data privacy and security policies and ensure that these policies are implemented across the organization. These policies should cover data retention, sharing, and breach response procedures.

Conducting Regular Security Audits And Risk Assessments; Regular security audits and risk assessments can help businesses to identify potential vulnerabilities in their CRM and take proactive steps to address these issues. Qualified professionals should conduct these assessments and include technical and procedural evaluations.

. . .

DATA PRIVACY and Security Concerns on CRM Success

If data privacy and security concerns are not adequately addressed, the success of the CRM may be compromised in several ways:

Reduced User Adoption; Employees may be reluctant to adopt and use the CRM if they do not trust that their and customers' data will be protected. This can lead to low user adoption, limiting the potential benefits of the CRM.

Loss Of Customer Trust; If customers do not trust that their data is being handled responsibly, they may be less likely to engage with the business and may even take it elsewhere. This can result in the loss of customers via reduced loyalty.

Legal And Financial Penalties; Failure to comply with data protection regulations can result in significant fines, legal liabilities, and reputational damage that can be difficult to recover.

Increased Risk Of Data Breaches and Cyberattacks; Without adequate privacy and security measures, the CRM may be more vulnerable, which can have severe consequences for the business.

You must prioritize privacy and security features to ensure your CRM is on the right track. This means investing time and resources into implementing robust access controls that restrict data access to only those who genuinely need it. By doing so, you can ensure that sensitive customer data remains secure and protected.

But it doesn't stop there. Employee training is just as crucial. Your team must know the importance of data privacy and security and understand the policies and procedures to safeguard customer information. Providing comprehensive training sessions will empower your employees to handle customer data responsibly and proactively address any concerns that may arise.

WARNING: The consequences of not addressing these concerns. Failure to prioritize data privacy and security can seriously affect your CRM. For starters, you may face reduced user adoption, which

means your employees and customers may be hesitant to embrace and utilize the system thoroughly. A CRM is only as good as the data people put into it.

BEYOND THAT, failure to address data privacy and security concerns will certainly lead to a loss of customer trust. Customers want to feel confident that their personal information is safe in your hands. If they sense any vulnerabilities, they may question your commitment to protecting their data and take their business elsewhere.

In today's digital landscape, cybercriminals constantly search for vulnerabilities to exploit. If your CRM has weak security measures (or you have not deployed and configured them to the fullest), you're rolling out the red carpet for these attackers. Trust me, dealing with a data breach, or cyberattack can devastate your business's reputation, finances, and legal obligations. Prioritize data privacy and security in your CRM. Implement strong access controls, conduct thorough employee training, and regularly assess and update your security protocols. By doing so, you can address potential concerns head-on and ensure the success of your CRM. Remember, the consequences of neglecting these factors are significant, but you can protect your business and build customer trust with the right approach.

21

Finding Saboteurs And Assholes

- EVERYONE HAS THEM 😱 -

A Catalyst for Organizational Change

When implementing a CRM system, it's crucial to recognize that change can be met with resistance, especially among certain individuals whom we may affectionately refer to as "difficult people" or even better "assholes." These individuals may harbor a fear of new technology, feel insecure about their job capabilities post-change, or staunchly cling to traditional methods. As experienced professionals, we know all too well that these disgruntled individuals can pose a significant barrier to the smooth adoption of a CRM.

WE KNOW that introducing a new system is never easy. Resistance to change can come from various sources, from technical difficulties to lack of training, and shifting established habits and methods can be challenging. However, one of the most significant obstacles is dealing with what some refer to as "difficult people." These individuals, for whatever reason, harbor skepticism, doubt, or outright resistance to the new CRM system.

It is crucial to identify and understand these "difficult people" in advance to steer through this challenge successfully. But how can organizations achieve this? Proactively gauging the workforce's composition enables management to empathize with these individuals' concerns, address their doubts and objections, and design a CRM adoption process that accounts for these issues accordingly. This strategic step showcases a deep understanding of the organizational landscape and lays the foundation for a smoother transition.

Planting the seeds of CRM adoption success begins with this fundamental insight. Organizations can foster a more collaborative environment by acknowledging the presence of these less-than-enthusiastic individuals and taking measures to mitigate their resistance. Companies can increase buy-in from critical stakeholders by investing time and resources in identifying these individuals and creating a CRM adoption process that accounts specifically for their needs.

However, the benefits of this approach go beyond merely increasing buy-in. It also lays the foundation for a smoother path to implementing and utilizing the CRM system. Dealing effectively with resistance to change minimizes the risk of encountering resistance or complications during the rollout. Addressing potential issues allows management to fine-tune and perfect the adoption process before implementation, consequently saving resources, preventing delays, and enhancing the project's overall outcome.

Identifying and understanding "difficult people" and taking measures to mitigate their resistance is a strategic step that you should implement before rolling out (preferably during the design phase) of a new CRM system. Doing so fosters collaboration, increases buy-in from critical stakeholders, and ultimately ensures the successful implementation and utilization of the CRM system.

Remember, acknowledging and mitigating ,even giving into if required, to potential challenges is not a sign of weakness but rather a hallmark of a strategic thinker who has been through it all. With your experience and wisdom, you can guide your efforts through this process and ensure a successful outcome.

Facilitates Strategic Planning

IN MY YEARS working with CRM systems, I've learned that change can sometimes trigger widespread and difficult-to-comprehend organizational resistance fully. We know that a particular group may struggle to adapt to new technology, while another may feel apprehensive about their capabilities post-change. At the other end of the spectrum, others may be adamant that traditional methods work best.

Identifying these potential resistors before launching a CRM is crucial as it can significantly impact strategic planning and execution. It allows organizations to prepare an actionable plan that caters to specific needs and sets realistic, measurable, and achievable goals.

By tailoring the plan to address potential roadblocks or obstacles, organizations can ensure that the change management process and training are more aligned with the actual needs of the workforce. For instance, if your team is technophobic, incorporating user-friendly guidance, comprehensive training, more generous periods for adaptation, and other similar resources into the launch plan can make the transition process much smoother and more aligned with organizational needs.

When implementing a CRM system, it's essential to take a pragmatic and cautious approach to change management. Detailed knowledge about potential opponents and their viewpoints not only renders the launch less disruptive but also minimizes the likelihood of backlashes down the line, thus ensuring the success of the CRM adoption in the long term.

Remember, a holistic and proactive perspective on these issues can help guide organizations through potential challenges and enable them to reap the most significant benefits from their CRM system. By drawing on experience, you can direct your individuals toward the critical importance of staying alert and proactively managing resistance throughout the process.

Enhanced Communication and Engagement

IN MY YEARS OF EXPERIENCE, I've encountered my fair share of difficult people during CRM implementations. These individuals can be roadblocks to the success of your CRM plans if their concerns and resistance are not effectively addressed. That's why developing enhanced communication plans tailored to engage and sway these problematic people is crucial, ultimately gaining their support.

Identifying the potential naysayers early on allows you to take a proactive approach to creating targeted communication strategies. You can design messages and initiatives that address their needs and alleviate their apprehension by understanding their specific objections, concerns, and motivations. The goal is to personalize communication to speak directly to their pain points and demonstrate how adopting the CRM system can address their challenges and benefit them personally.

To begin, you must gather insights about these problematic individuals. **Engage with them through one-on-one discussions**, focus groups, or surveys to understand their perspectives, motivations, and reasons behind their resistance. This information will help you tailor your messaging and communications to resonate with them more profoundly.

Once armed with this valuable data, it's time to develop enhanced communication plans that target their concerns directly. **Craft messages that clearly articulate the benefits** of the CRM system and how it aligns with their own goals and objectives. Use relatable examples and success stories from similar colleagues or departments to showcase the positive impact the CRM can have on their work.

It's important to **communicate early and often, using a variety of channels** to reach these difficult people effectively. This may include regular email updates, town hall meetings, training sessions, and individual coaching or mentoring if necessary.

The key is **consistently providing information**, addressing concerns, and actively seeking their input throughout the process.

Special attention may sometimes be required for the most challenging individuals. For those resistant due to a lack of understanding or fear of change, consider offering **focused training sessions or workshops** that address their specific knowledge gaps and provide hands-on practice with the CRM system. Providing them with the **comfort, tools, and resources** they need to succeed will help alleviate their fears and build confidence in the new system.

Remember, swaying and gaining the support of difficult people is not an overnight task. It requires **patience, empathy, and a willingness** to address their concerns head-on. Be prepared to **listen actively and adapt** your communication strategies as needed. Continuous dialogue and feedback loops are essential to monitor progress, address further objections, and keep these individuals engaged.

By investing time and effort into targeted communication plans, you can actively engage complex individuals and turn them into advocates for the CRM system. **Their support can significantly impact the success of the implementation and help minimize roadblocks along the way**. However, it's important to note that not all difficult people may be won over, and in such cases, it's crucial to have mitigation strategies in place to manage their impact on the broader CRM plans.

A CRM system must enhance communication plans to engage difficult people. By understanding their concerns and motivations, crafting personalized messages, and providing ongoing support, you can sway these individuals and gain valuable support for the CRM initiative. Remain persistent and adaptable, and with due diligence, and you can overcome their resistance and achieve CRM success.

Mitigate Resistance and Efficient Change Management

I HAVE ENCOUNTERED my fair share of difficult individuals who can challenge change management. These individuals can be a significant roadblock to the success of your CRM plans if their resistance is not effectively addressed. However, with proactive identification and targeted action, you can turn them into valuable allies and gain their support for the implementation.

Resistance to change is a common challenge and a significant contributing factor to the high failure rate of CRM implementations. Identifying these resistors early on is crucial to proactively manage their resistance rather than trying to reactively control the situation once the CRM system is in place.

Understanding the reasons behind the resistance and the individuals involved is the first step in effectively mitigating their opposition. Knowing who these difficult people are, you can tailor your strategies and actions to address and neutralize their objections. This proactive approach allows you to integrate steps into your change management process that specifically target their concerns and motivations.

One effective strategy is to involve these resistant individuals in decision-making processes or the early testing phases of the CRM system. This engagement enables them to gain firsthand experience and familiarity with the system, giving them a sense of ownership and control. When assholes feel like they have a say in the process, they are more likely to reduce their initial resistance and become more accepting of the changes ahead.

Moreover, it is essential to communicate openly and transparently with these individuals, keeping them informed and involved throughout the implementation process. Be prepared to address their concerns head-on, providing clear and honest answers. You can build trust and rapport with even the most resistant assholes by understanding their objections and respectfully addressing them.

Remember, the key to swaying these individuals lies in

demonstrating the benefits and value of the CRM system to their work and success. This is due to the fact that they are very likely professionally selfish. Tailor your messaging to highlight how the implementation aligns with their goals and objectives, showcasing how it can make their jobs more accessible and efficient and ultimately improve their results. Provide them with tangible examples and success stories from similar colleagues or departments who have already experienced the positive impact of the CRM system.

Incorporating ongoing training and support for these individuals is also essential. Offer specialized training sessions that address their needs and concerns, ensuring they have the knowledge and skills to navigate the CRM system effectively. Provide them with continuous guidance and mentorship to help them overcome any challenges they may face during the transition. This level of investment shows your commitment to their success and can further foster their acceptance of the changes.

Identifying and effectively managing assholes resistant to change is crucial for successful CRM implementation. By proactively targeting their concerns, involving them in decision-making, and providing ongoing support, you can turn these difficult individuals into valuable allies who support and actively champion the implementation. Remember, this process requires patience, empathy, and a willingness to address their objections with the benefit of your experience. You can mitigate resistance and foster efficient organizational change management with due diligence and strategic action.

The CRM can be under-utilized if internal roadblocks due to resistant individuals are not identified and tackled before the launch. This can derail the purpose of implementing such a system to improve customer relationships, streamline processes, increase productivity, and drive profitability.

Early identification of such personnel enables the design of user-specific training programs, tackles resistance, and ensures the CRM's maximum utilization in the organization while driving the expected value.

Difficult People's Identity 🔍

BY IDENTIFYING and addressing the challenges posed by saboteurs, assholes, and sticks-in-the-mud, organizations can mitigate their impact on the CRM's success and convert them into supporters. This will ultimately lead to a more corporate-aligned and efficient CRM, benefiting the entire organization.

Observe Their Behavior; Pay close attention to the actions and attitudes of internal leaders during meetings, workshops, and other project-related activities. Watch for signs of resistance, such as unwillingness to engage in discussions, reluctance to share information, or consistently negative attitudes toward the CRM project.

Gather Feedback From Team Members; Seek input from other team members about their experiences working with these leaders. They may provide valuable insights into the leaders' behavior and attitudes, helping you identify those resistant to the CRM project.

Analyze Communication Patterns; Review email correspondence, meeting minutes, and other forms of communication to identify leaders who consistently express concerns or raise objections to the CRM project. This may indicate a pattern of resistance and unwillingness to cooperate.

Assess Their Influence; Consider the level of influence these leaders have within the organization. Those with significant decision-making power or control over resources may pose a greater challenge to the CRM project if they resist change.

Below are a few types of difficult individuals who may impede the success of a CRM, along with ways to identify them, the potential harm they could cause, and the consequences of not addressing these issues.

The Resistant Traditionalist

HOW TO IDENTIFY: This individual may hesitate or resist adopting new technologies and processes, often preferring to stick with their familiar working methods. They may express concerns about the complexity of the CRM or doubt its ability to improve their work.

Potential harm: Their resistance to change can slow the implementation process, create a hostile atmosphere among team members, and limit the organization's ability to adapt and innovate.

Cons of doing nothing: If their concerns are not addressed, they may resist the CRM, potentially influencing others to do the same and undermining its effectiveness.

Strategies for conversion:

Provide comprehensive training and support to help them understand the benefits and features of the CRM. Address their concerns by showcasing how the CRM can improve efficiency, streamline processes, and simplify their job. Involve them in the decision-making process and invite them to provide feedback and suggestions for improvement.

The Overwhelmed Skeptic

HOW TO IDENTIFY: This person may feel overwhelmed by the CRM's features and capabilities, leading them to doubt its usefulness or question its value. They may be vocal about their concerns and hesitant to invest time and effort into learning the new system.

Potential harm: Their skepticism can create doubt among other team members, delay the adoption of the CRM, and hinder the organization's ability to achieve its goals.

Cons of doing nothing: If their concerns are not addressed, they may continue to question the value of the CRM, potentially causing delays in its implementation and adoption.

Strategies for conversion:

Offer personalized training and support to help them build confidence using the CRM. Break down the CRM's features and benefits into manageable components, allowing them to see the value in each aspect. Share success stories and examples of how the CRM has improved processes and outcomes for other organizations or teams.

The Passive-Aggressive Non-User

HOW TO IDENTIFY: This individual may outwardly agree to use the CRM but may not adopt it daily. They may find ways to work around the system or use it minimally, often citing a lack of time or resources as their reason.

Potential harm: Their lack of engagement with the CRM can result in incomplete or inaccurate data, reduced efficiency, and a lack of collaboration among team members.

Cons of doing nothing: If their passive-aggressive behavior is not addressed, they may avoid using the CRM, undermining its effectiveness and creating additional work for other team members.

Strategies for conversion:

Set clear expectations and deadlines for completing CRM-related tasks and hold them accountable for meeting these requirements. Offer support and resources to help them overcome any CRM barriers, such as time constraints or a lack of understanding. Recognize and reward their progress in adopting the CRM, reinforcing the value of their efforts and encouraging continued engagement.

The Misaligned Manager

HOW TO IDENTIFY: This individual may be a manager or team leader who does not fully understand or support the CRM's objectives. They may prioritize their department's goals over the organization's overall goals or resist the changes brought about by the CRM.

Potential harm: Their misaligned priorities can create confusion, slow the implementation process, and hinder the organization's ability to achieve its goals.

Cons of doing nothing: If their misaligned priorities are not addressed, they may continue to resist the CRM, potentially causing delays in its implementation and adoption and negatively impacting the organization's overall performance.

Strategies for conversion:

Communicate the organization's goals and objectives and demonstrate how the CRM aligns with these goals. Involve them in the planning and implementation process, allowing them to see the value of the CRM firsthand and understand its benefits for their department. Provide opportunities for growth and development, such as training and workshops, to help them adapt to the new CRM and align their priorities with the organization's goals.

The Uncooperative Team Member

HOW TO IDENTIFY: This person may be uncooperative or unresponsive when using the CRM, often failing to complete tasks, attend training sessions, or provide necessary information.

Potential harm:_Their lack of cooperation can result in incomplete or inaccurate data, reduced efficiency, and a lack of collaboration among team members.

Cons of doing nothing: If their uncooperative behavior is not addressed, they may continue to hinder the CRM's progress, negatively impacting its effectiveness and creating additional work for other team members.

Strategies for conversion:

Set clear expectations and deadlines for completing CRM-related tasks and hold them accountable for meeting these requirements. Offer support and resources to help them overcome CRM barriers, such as a lack of understanding or time constraints. Foster a culture of teamwork and collaboration, encouraging open communication and shared responsibility for the CRM's success.

The Know-it-All

HOW TO IDENTIFY: This person is characteristically confident about their knowledge of CRMs. They may resist formal training or dismiss the new system because they believe they already understand everything they need to.

Potential harm: Their overconfidence may lead to overlooked features or misuse, undermining the CRM's utility and potential capabilities.

Cons of doing nothing: If such behavior is not addressed, they can negatively influence other team members and create misunderstandings about the CRM.

Strategies for conversion:

Involve them in the CRM training process, converting their knowledge into a productive influence for the team. Encourage them to showcase the proper use of the system, integrating their experience into the team's learning process. Leverage their knowledge to motivate other members, bridging the gap between experienced and beginner-level associates.

The Noncommittal Observer

HOW TO IDENTIFY: They may cautiously observe the CRM and its effects on the team without actively engaging. Their commitment level may be low due to a lack of understanding or low confidence.

Potential harm: Their reserved approach might hinder effective team collaboration and result in a sluggish CRM adoption process.

Cons of doing nothing: If never fully committed, they are a hurdle to CRM success, causing team disunity and lowered productivity.

Strategies for conversion:

Encourage active participation in CRM-related discussions and training to boost their confidence. Assign CRM tasks to them in stages, increasing their involvement gradually over time. Provide continuous feedback on their CRM-related performance to help them understand their importance for the system's success.

I HAVE SEEN organizations fall into the trap of underestimating the impact of challenging individuals on the CRM implementation process. They can derail progress, create conflicts, and even lead to project complete failure if not handled properly. By heeding this cautionary advice and investing the necessary time and effort into addressing these challenges head-on, organizations can pave the way for a truly successful CRM implementation that drives efficiency, cultivates strong customer relationships, and propels overall success for years to come. Remember, the path to CRM success is not always easy. Still, with the right approach and the benefit of knowledge gained from real-life experiences, you can navigate these challenges and reap the rewards that a well-implemented CRM system can bring.

BELOW ARE some more even **highly toxic** types of difficult individuals that may impede the success of a CRM and ways to identify them:

The Saboteur "Mega Asshole"

THIS INDIVIDUAL MAY INTENTIONALLY UNDERMINE the CRM's success by providing false or misleading information, deleting or tampering with data, or creating unnecessary obstacles. They might do this due to personal grievances, fear of change, or to protect their interests.

How to identify; Look for patterns of behavior that seem counterproductive or harmful to the CRM's success. This may include spreading negative rumors, consistently providing inaccurate data, or actively resisting the implementation of new processes.

Ideas To Convert Them; Open communication via Establishing a dialogue with the individual to understand their concerns and motivations. Address any misconceptions or fears they may have about the CRM. Involve them in the process by giving them a role in implementing or managing the CRM, allowing them to feel more invested in its success. Monitor their actions by keeping a close eye on their activities and address any harmful behaviors immediately.

The Naysayer "Disrespectful Asshole"

THIS PERSON IS CONSISTENTLY negative about the CRM, often expressing doubts about its effectiveness, questioning its value, or predicting its failure. Their negativity may stem from a fear of change, a lack of understanding of the system's benefits, or a desire to maintain the status quo.

How to identify: Respond to individuals who frequently express pessimistic views about the CRM or resist adopting new processes and technologies. They may be vocal in meetings, share their negative opinions with others, or be uncooperative during training sessions.

Ideas To Convert Them: Provide education and training to ensure they thoroughly understand the CRM's benefits and capabilities through training sessions and educational materials. Share success stories by highlighting examples of how the CRM has improved processes, increased efficiency, or benefited other team members. Address their concerns and listen to their objections and address them with factual information and positive outcomes.

The Manipulator "Conniving Asshole"

THIS INDIVIDUAL MAY USE the CRM to their advantage by manipulating data or processes to benefit themselves or their department. They may withhold crucial information, exaggerate results, or misrepresent data to make themselves look better or to gain an unfair advantage.

How to identify: Look for discrepancies in data or unusual patterns that suggest manipulation. This may include data that consistently favors a specific individual or department or instances where the CRM is being used in ways that deviate from its intended purpose.

Ideas To Convert Them: Establish clear rules and guidelines and implement strict policies regarding data integrity and CRM usage to discourage manipulation. Monitor data and processes by regularly audit the CRM for signs of manipulation and address any discrepancies immediately. Encourage transparency and accountability by fostering a culture of openness and honesty, where team members are encouraged to share information and collaborate.

The Power Player "Faux Strategic Asshole"

THIS PERSON MAY SEE the CRM as a threat to their position of power or influence within the organization. They may resist the implementation of the system or attempt to control its usage by restricting access, limiting training, or dictating how the system should be used.

How to identify: Watch for individuals who seem overly protective of their authority or resist sharing information and resources. They may be reluctant to delegate tasks, insist on controlling access to the CRM, or make unilateral decisions about its usage without consulting others.

Ideas To Convert Them: Demonstrate the benefits of the CRM by showing how it can enhance its position by improving efficiency, providing valuable insights, and streamlining processes. Encourage collaboration by fostering a culture of teamwork and cooperation, where all team members have a say in decision-making and are encouraged to contribute their ideas. Provide growth opportunities by offering training and development opportunities to help them expand their skill set and adapt to the new CRM.

The Procrastinator "Lazy Asshole"

THIS INDIVIDUAL MAY DELAY or avoid using the CRM, often citing a lack of time, resources, or understanding. Their procrastination may be due to a fear of change, a lack of motivation, or a desire to avoid accountability.

How to identify: Look for individuals who consistently miss deadlines, avoid participating in training sessions, or fail to complete tasks related to the CRM. They may make excuses for their lack of progress or claim they are too busy with other responsibilities.

They may make excuses for their lack of progress or claim they are too busy with other responsibilities.

Ideas To Convert Them: Set clear expectations and deadlines: Establish specific goals and timelines for completing CRM-related tasks and hold them accountable for meeting these deadlines. Offer support and resources: Provide the necessary tools, training, and assistance to help them overcome any CRM barriers. Recognize and reward progress: Acknowledge their achievements and progress in adopting the CRM, reinforcing the value of their efforts.

Winning People Over

THE SUCCESS or failure of a newly-deployed CRM hinge mainly on the commitment, support, and transparency of an organization's internal leaders. Their honesty regarding their concerns and needs during the post-deployment phase is paramount, especially when those leaders demonstrate resistance to change, reluctance to share vital information or behaviors that may create obstacles in the project's progress.

A CRM'S purpose revolves around streamlining business operations, enhancing customer relationships, and driving organizational success. Recognition of the integral role of internal leaders in realizing these objectives becomes even more acute, considering their influential positions and ability to sway decisions within the organization.

THE INITIAL BARRIER TO CRM deployment success is typically resistance to change, a prevalent human trait profound within individuals in significant decision-making roles. Change implies stepping out of comfort zones, adopting new paradigms, and embracing new methodologies – tasks that can be daunting for some. When exhibited by crucial internal leaders, this resistance poses substantial threats to the full-fledged adoption of CRM.

RENUNCIATION OF CHANGE among these leaders can go to lengths to delay the assimilation and acceptance of the CRM within the organization. This delay curbs the optimal flow and functionality of the CRM, affecting its immediate benefits, such as improved productivity, decreased operational costs, and heightened levels of efficiency. It also gives rise to a pervasive domino effect—creating an impediment to transformation, slowing down the synchronization of business processes with the CRM, and subverting potential advantages the system could otherwise provide.

INTERNAL LEADERS' reluctance to share information poses yet another hurdle post-CRM deployment. A CRM demands consistent data inputs for maintaining data quality, accuracy, and reliability—fundamental elements to ensure informed decision-making, precise forecasting, and personalized customer interaction. Non-compliance in sharing information not only encapsulates a potential risk of derailing the CRM project but also delineates a risk of negatively affecting the organization's direction, altogether disrupting the decision-making process.

HOWEVER, the stakes are even higher when internal leaders exhibit obstructive behavior toward the project's progress following the CRM's deployment. Such behavior can not only thwart the overall effectiveness of the project implementation but also dampen the morale of other team members. Skepticism, doubts, and uncertainties can rapidly permeate the organization's ranks, jeopardizing the system's acceptance and success. This potential hurdle, if ignored, can translate into a profound cultural deficit across the organization, fostering a negatively-skewed perception of the CRM.

WHEN REMEDIATION EFFORTS aren't deployed efficiently to address these issues, the organization's failure to realize the benefits of CRM is inevitable. Delayed adoption caused by change resistance or withholding information can manifest in a drawn-out transition phase punctuated by inefficiency and stunted growth. A CRM deployment without timely remediation measures potentially results in less cohesion among members, ultimately creating distrust and resistance.

. . .

FURTHERMORE, failure to effectively remedy these problems affects the ROI of the CRM. Limited system usage contributes to reduced productivity, longer sales cycles, less efficient marketing efforts, and, ultimately, reduced customer satisfaction. In terms of financial implications, the organization may struggle to see the quantifiable benefits of CRM due to a lack of full utilization.

In worst cases, where stubborn obstruction to the project's progress prevails, complete failure of CRM implementation isn't a far-fetched scenario. The inability to remedy a toxic culture of non-cooperation could eventually lead the organization to revert to original, often outdated business processes before deploying the CRM, negating the imagined benefits of organizational efficiency and customer relationship enhancement. This reversion results in the squandering of invested resources, wasted efforts, and a pronounced dip in overall corporate morale.

ANY COMPROMISE on customer satisfaction due to these challenges negatively impacts business relationships, customer retention rates, and, ultimately, the bottom line. This failure defines the line between garnering a competitive edge in the market and facing a setback in losing customer loyalty, innovation, and business growth.

THE ABSENCE of timely remediation following the detection of issues in post-CRM deployment can lead to dire consequences. The ramifications are significant, from a stall in CRM adoption, compromised data integrity, and estranged organizational culture to a possible failure of the CRM implementation project altogether.

POST-CRM DEPLOYMENT, successful CRM implementation depends extensively on a problem-solving approach where potential issues are identified and acted upon efficiently. Remediation, therefore, ceases to be just an option - it becomes an imperative process that mitigates the risk of disastrous outcomes, ensuring organizational success and providing the foundation for the maximized utilization of the CRM. The stakes of not administering remediation post-identification are high, featuring a domino effect on organizational growth, morale, and customer satisfaction, underlining the critical role remediation plays in post-CRM deployment.

HERE ARE some ideas that may (or not) work in your situation to win people/teams/divisions/geographies over:

- **Build personal relationships:** Establishing rapport and trust with difficult leaders is essential for overcoming their resistance. Schedule one-on-one meetings, invite them to lunch, or engage in informal conversations to better understand their concerns and motivations. Building personal relationships can create a foundation for open communication and collaboration.

- **Understand their concerns and needs:** Listen attentively to the concerns and needs of resistant leaders, and try to empathize with their perspective. This will help you identify the causes of their resistance and develop tailored strategies to address their specific concerns.

- **Present the value proposition:** Clearly articulate the benefits of the CRM project for the organization and individual leaders. Highlight how the project can help them achieve their personal and professional goals, such as increased efficiency, improved decision-making, or enhanced customer relationships.

- **Leverage their expertise:** Involve resistant leaders in the CRM project by assigning them specific roles or responsibilities that capitalize on their unique skills and expertise. This can help them feel valued and invested in the project's success, ultimately increasing their willingness to cooperate.

- **Provide education and training**: Offer targeted education and training sessions to address the concerns and needs of resistant leaders. This can help to dispel misconceptions, build confidence in the CRM, and demonstrate your commitment to addressing their concerns.

- **Facilitate cross-functional collaboration:** Encourage collaboration between difficult leaders and other team members by organizing workshops, brainstorming sessions, or joint projects. This can help to break down silos, foster a sense of teamwork, and create opportunities for resistant leaders to see the benefits of the CRM project firsthand.

- **Highlight success stories:** Share examples of other organizations or departments within your company that have successfully implemented CRMs and experienced positive results. This can demonstrate the potential benefits of the CRM project and inspire resistant leaders to reconsider their stance.

- **Create a sense of urgency:** Advertise and promote the importance of the CRM project for the organization's long-term success and the potential consequences of not adopting the system. This can help to create a sense of urgency and motivate resistant leaders to take action.

- **Offer incentives:** Consider offering incentives, such as bonuses, promotions, or public recognition, to leaders who actively support the CRM project and are willing to change their behavior. This can create a positive reinforcement loop, encouraging resistant leaders to cooperate more.

- **Be persistent and patient:** Winning over difficult internal leaders may require time, persistence, and patience. Continuously engage with these individuals, address their concerns, and demonstrate your commitment to the CRM project's success. Over time, their resistance may diminish, and they may become more willing to cooperate.

- **Leverage peer influence:** Identify influential peers within the organization who support the CRM project and advocate for its benefits. Encourage these individuals to engage with resistant leaders and share their positive experiences, helping to sway their opinions.

- **Manage expectations:** Communicate the goals, objectives, and timelines of the CRM project to resistant leaders, and manage their expectations throughout the process. This can help prevent misunderstandings and show leaders know the project's progress and potential challenges.

- **Address concerns transparently:** When resistant leaders raise concerns or objections, address them transparently and honestly. This can demonstrate your commitment to addressing their concerns and creating an environment of openness.

- **Demonstrate flexibility:** Show a willingness to adapt the CRM project plan based on the feedback and concerns of resistant leaders. This can demonstrate that their input is valued and encourage them to cooperate more.

- **Lead by example:** As the CIO, model the behavior and attitudes you expect from internal leaders. Demonstrate your commitment to the CRM project, actively seek feedback and be responsive to the concerns and needs of all stakeholders. Leading by example can set the tone for the organization and encourage others to follow suit.

ADDRESSING the challenges posed by challenging and important internal leaders is crucial to the success of our CRM project. By identifying these individuals and employing creative techniques to win them over, we can foster a more collaborative and supportive environment, ultimately ensuring the project's success.

It's True; Humans Matter 🤪

IN THE END, difficult or easy, understanding and embracing all of the people in the organization is critical.

The human aspect in a CRM launch is a pivotal factor for a successful transformation. It's essential to understand that managing 'difficult' individuals isn't simply about jumping hurdles.

Instead, it's enriching the positive influence on CRM adoption. Identifying these personas within the organization before the CRM launch forms a strategic advantage - ensuring a smoother implementation process and a greater probability of the CRM's success in the organization. This human-aware, considerate approach nurtures a collaborative, efficient, and effective techno-cultural environment that drives individual and organizational growth.

22

Case Study: Performers, Champions, And Sponsors

Scene:

You're an experienced CRM deployment manager, having seen many successful and unsuccessful deployments. As you embark on a new CRM deployment project, you recall a critical lesson: the importance of directly accessing different levels of people from each team, division, or geography using the CRM platform.

Incident:

During the process of having regular meetings with the core team, it has become apparent that specific discussion points are "leaking" outside of the efforts the CRM core team liaison meets with representatives from each group. Additionally, the tone of the meetings is not as friendly as it once was, and you, as the leader, are hearing more concerns during a typical day (outside of the core meetings). It becomes evident that some individuals' motivations and competencies are questionable.

While you would like to believe that everyone is attending for good business development purposes and is genuinely interested in the technology and how it can benefit their group, some people have other reasons for attending.

Result:

The people involved in the CRM deployment can make or break the project's success. The project can fail when competing teams have sent representative liaisons or are professionals not contributing efficiently. Individuals with spare time or sunsetting careers may not have the drive to ensure successful deployment. Additionally, the developer of group-specific offline tools may resist transitioning to the more extensive CRM, leading to delays, setbacks, and lost productivity.

Lessons Learned:

Ensure Motivation; Carefully evaluate who will represent each group during the CRM deployment. Verify their motivations, seek out people with a genuine interest in the project's development, and establish clear expectations of participants from the start.

Confirm Competency; Ensure that liaisons assigned to the CRM core team are competent and understand how to use the CRM platform. Invest in proper training to ensure everyone understands the system, its requirements, and the necessary involvement for a successful CRM deployment.

Plan Appropriate Incentives; Encourage your team throughout the deployment. Having professional and personal incentives to motivate people is critical to ensuring everyone contributes to the project's success.

What Should Have Been Done Differently?

To avoid unfavorable outcomes related to competency and motivation, it is essential to:

Conduct a Comprehensive Selection Process; Conduct a thorough approval process for representatives from each group, verifying their motivations and credentials beyond just nominating individuals based on their current roles.

Prioritize Clear Expectations; Establish detailed expectations of all liaisons and their roles before deployment. Clear communication is necessary to ensure everyone is in sync and working towards the same goals.

Communicate Benefits to All; Communicate how the CRM will enhance the productivity and competitiveness of each group. Addressing the inherent tendency of group members to cling to their current ways of working is crucial to the successful promotion and adoption of CRM.

WHILE DIRECT ACCESS to people from each team, division, or geography is critical to any successful deployment and operation of a CRM, it is equally imperative to have competent and motivated individuals involved.

Don't be so thankful to receive just anyone to the core team - Careful selection and preparation of liaisons, verification of their competence, clear communication, and incentivization will all support that a CRM deployment improves sales performance across the entire organization, and each team views the CRM deployment as a business advantage.

Part III: Optimizing The Operations

In this book's third and final section, we embark on a journey to explore crucial topics that often arise when an organization is still grappling with the shock of implementing an official (universal) CRM. During this phase, we must adhere to the principle of due diligence, considering the concepts discussed thus far and drawing insights from our unique circumstances. This is the key takeaway: the areas we delve into extend far beyond the setup phase alone. Instead, they can emerge throughout the entire CRM lifecycle, from the initial implementation to the ongoing maintenance and future upgrades.

Equipping ourselves with a deep understanding of these topics early on is paramount. Doing so allows us to fortify our position and better prepare ourselves for unexpected challenges that may ensue down the line. Our goal is clear: we must proactively identify and address potential issues, thus enabling us to combat any situation as it arises.

To successfully navigate this phase, close attention must be paid to critical areas where challenges may manifest. It is essential to remember that you are not alone in facing these circumstances. Numerous professionals who came before you have encountered similar challenges and successfully resolved them. Drawing from their experiences allows us to navigate the CRM landscape more confidently and effectively.

By thoroughly exploring and addressing these critical areas, you will be well-prepared to navigate the CRM challenges that may arise during its lifecycle. As you progress through this section, remember that the insights and lessons shared with you are born out of real-world experiences and the successful remedies employed by professionals like yourself.

Remain diligent, stay prepared, and let the knowledge gained from others' experiences guide you toward CRM success.

1

Measuring CRM Success

A CRM system is a powerful tool for businesses to manage customer interactions, streamline processes, and improve overall customer experience. A critical aspect of making a CRM successful is identifying, presenting, and being transparent about the key performance indicators (KPIs) and other metrics required by management.

Importance of Identifying and Presenting KPIs and Metrics

Effective identification and presentation of KPIs and metrics in a CRM are crucial for several reasons:

Informed Decision-Making; As an experienced professional in CRM, I cannot stress enough the importance of informed decision-making for executives and other leaders in any organization. Accurate and relevant key performance indicators (KPIs) and metrics are integral to the decision-making process when managing sales, marketing, customer service, and product development strategies. However, this is not limited to just the leaders of an organization but rather a continual issue faced by project teams and administrators as well.

Let us explore some real-world examples to illustrate how informed decision-making works in different areas of an organization. Sales managers can gain insights into their team's performance by tracking KPIs such as conversion rates, average deal size, and sales velocity. For instance, declining conversion rates may signal a need for further analysis to understand why and the appropriate actions, such as additional training or revision of the sales process, can be taken. On the other hand, if the average deal size is trending upward, it may suggest that the team is successfully upselling or cross-selling, making it necessary to adjust sales strategies to capitalize on these opportunities.

Marketing efforts can be precisely analyzed and optimized using KPIs and metrics, enabling marketers to understand the effectiveness of their campaigns. By tracking website traffic, conversion rates, and cost per lead, marketers can better understand which marketing channels generate more information at a lower price, prompting a reallocation of resources. Conversely, campaigns that are not performing well can be identified, and data-driven adjustments made to increase their impact.

Another critical area in CRM is customer service. KPIs like customer satisfaction scores, response time, and first-contact resolution rate can provide valuable insights into the effectiveness of a company's customer service efforts. For instance, consistently low customer satisfaction scores may highlight a need for additional training or process improvements. Organizations can proactively address joint pain points through metrics and enhance the customer experience by identifying common pain points.

Product development KPIs related to product performance, customer feedback, and time-to-market can guide decision-making. By monitoring customer feedback and usage data, organizations can see areas for improvement and prioritize product enhancements or new feature developments. Metrics such as time-to-market can help evaluate the effectiveness and efficiency of development processes, enabling teams to make better decisions on resource allocation and project timelines.

The ability of executives and other leaders in a company to make informed decisions based on accurate and relevant KPIs and metrics is of the utmost importance. However, this is not just limited to executives but is a continual issue faced by project teams and administrators. By having these vital components in place, an organization can operate more effectively and reach its goals more efficiently. Though it is important to highlight potential pitfalls by having a cautionary tone, due diligence should always be an integral part of decision-making processes in CRM.

Performance Evaluation; One key aspect is tracking customer acquisition costs. By analyzing the cost-effectiveness of sales and marketing efforts, businesses can optimize their allocation of resources and maximize their return on investment. By monitoring the expenses incurred in acquiring new customers, companies can make informed decisions about the most effective channels and strategies for attracting and converting leads. This data allows teams to refine their approach, streamline processes, and focus on activities yielding the highest value and conversion rates.

Another important metric for performance evaluation is measuring customer lifetime value. This metric allows businesses to assess the long-term profitability of their customer base. By evaluating the profitability of individual customers over their entire relationship with the company, businesses can identify high-value customers and tailor their strategies to maximize the profitability of these relationships. By understanding the long-term value of customers, companies can make strategic decisions around customer segmentation, loyalty programs, and customer retention initiatives.

Email campaigns play a crucial role in customer engagement and marketing efforts. Evaluating the response rates of email campaigns is a key performance metric that helps businesses optimize their messaging and increase customer engagement. Companies can gain insights into the effectiveness of their messages, design, and targeting by analyzing open rates, click-through rates, and conversion rates from email campaigns. This data allows teams to refine email marketing strategies, personalize content, and improve customer engagement and conversion rates.

Customer complaints or escalations serve as valuable feedback for businesses. By measuring the number of customer complaints or escalations, companies can identify areas for improvement in service quality. This metric allows enterprises to detect pain points in their customer journey, identify recurring issues, and take proactive steps to address them. Businesses can enhance customer satisfaction, loyalty, and advocacy by actively listening to customer feedback and resolving issues promptly. Managing customer complaints improves customer experience, safeguards the brand reputation, and strengthens customer relationships.

Tracking the rate of <u>customer retention or churn</u> is a crucial metric for reviewing the effectiveness of customer retention strategies. Retaining existing customers is almost always more cost-effective than acquiring new ones (not to further detail the likelihood of ill-will and negative social media comments), and measuring customer retention rates is essential for understanding the success of customer retention initiatives. By monitoring churn rates, businesses can find patterns that may indicate issues with the product, service, or overall customer experience. This data enables organizations to implement targeted retention strategies, such as loyalty programs, personalized communication, and proactive customer support, to reduce customer churn and increase customer loyalty.

It is important to emphasize that performance evaluation is a continual process that requires the active involvement and commitment of leaders, project teams, and administrators. Regularly reviewing and analyzing performance metrics in collaboration with all stakeholders is essential for achieving desired outcomes. The insights from performance evaluation enable businesses to make data-driven decisions, adjust strategies, and drive continuous improvement.

Implementing performance evaluation using KPIs and metrics is crucial for measuring the success of CRM initiatives and driving business growth. By tracking customer acquisition costs, measuring customer lifetime value, evaluating email campaign response rates, monitoring customer complaints or escalations, and tracking customer retention or churn rates, businesses can gain valuable insights to optimize their strategies, improve customer experiences, and enhance overall business performance. Continual evaluation and refinement are necessary to adapt to changing market dynamics, customer expectations, and emerging trends. By embracing performance evaluation practices, businesses can stay ahead of the competition, drive sustainable growth, and foster long-term customer relationships.

Resource Allocation; The availability and allocation of resources are crucial to support a CRM system's implementation, maintenance, and ongoing optimization. Whether it is financial resources, human resources, or technological resources, aligning and ensuring the proper resources are in place is vital for the success of any CRM initiative.

In marketing campaigns, tracking KPIs is essential to measure the effectiveness and success of various initiatives. Management gains valuable insights into which marketing campaigns perform best by meticulously monitoring metrics such as conversion rates, customer acquisition costs, and return on investment (ROI). This enables them to allocate more resources to these successful campaigns, optimizing the marketing budget and achieving the highest possible return on investment. It also allows for reallocating resources away from underperforming initiatives, ensuring that resources are focused on efforts that yield the most significant results.

Product development is another area where KPIs and metrics are crucial in measuring CRM success. Regarding multiple projects in progress, decision-makers must prioritize resource allocation based on potential revenue or market demand. By identifying KPIs such as market size, customer feedback, and expected profitability, management can make informed decisions on where to allocate resources effectively. For example, this could mean directing more resources to products or features with higher market demand or adjusting resource allocation based on the potential revenue each project can generate. This approach ensures that resources are utilized efficiently, maximizing the chances of success for each task and ultimately contributing to the overall success of the CRM system.

Effective resource allocation is critical for the success of sales teams operating in different territories or regions. Identifying and tracking KPIs such as sales growth rate, revenue per territory, and customer acquisition costs can catalyze transformative change. Management can strategically allocate resources by understanding each sales territory's performance through these metrics. If a region performs exceptionally well and has high growth potential, allocating additional salespeople or marketing resources can further accelerate its success. Alternatively, adjustments can be made if a territory is underperforming, such as providing extra support or training to improve results. Carefully allocating resources to sales territories enables organizations to optimize their sales efforts and drive CRM success.

Efficiently allocating customer service resources is crucial in handling customer inquiries promptly and effectively. Tracking metrics such as response time, resolution rate, and customer satisfaction scores provides management with valuable insights into areas that need improvement. With this data in hand, resources can be allocated accordingly. For example, if customer service response times are consistently high, reallocating resources to increase staffing or implementing technology solutions like chatbots can help address this issue and improve customer service performance. By aligning resources with the areas that need improvement the most, organizations can enhance their customer service delivery and contribute to the overall success of the CRM system.

While using KPIs and metrics is undoubtedly crucial in measuring the success of a CRM system, it is essential to approach these tools with caution and diligence. Mindlessly relying on numbers without considering the broader context can lead to misguided decisions that may have far-reaching consequences. It is crucial to ensure the quality and relevance of the metrics being tracked and to interpret the data with a comprehensive understanding of the situation at hand. This requires ongoing evaluation and adjustment to ensure that the selected KPIs and metrics genuinely reflect the goals, objectives, and unique characteristics of the CRM system and the organization.

The allocation of resources in various areas, such as marketing campaigns, product development, sales territories, and customer service, plays a crucial role in driving CRM success.

Goal Setting And Tracking; Goal setting and tracking are essential components of measuring the success of a CRM system. One critical area for goal setting and tracking in CRM is revenue growth. Many businesses set goals to increase annual revenue; evaluating progress toward this objective is vital. Management can gain insights into their revenue growth trajectory by tracking critical financial metrics like revenue growth rate, average deal size, and customer profitability. If the revenue growth rate falls below the desired target, it may indicate the need to make strategic adjustments. For example, increasing marketing efforts, expanding into new markets, or refining the sales process may be necessary to stimulate revenue growth. By regularly monitoring these metrics, management can identify trends, make data-driven decisions, and take proactive measures to achieve revenue growth goals.

Customer retention is another crucial factor in measuring CRM success. Improving customer retention and reducing customer churn is essential for long-term business success. KPIs such as customer churn rate, customer satisfaction scores, and repeat purchase rates are invaluable tools for monitoring progress. High customer churn rates can indicate other issues that need to be addressed, such as poor customer service or product dissatisfaction. By tracking these metrics and assessing the customer experience, management can identify areas for improvement and take the necessary actions to retain more customers. With a better understanding of your customer's behavior and preferences, adjustments can be made to CRM strategies and initiatives to enhance customer retention and loyalty.

In addition to tracking organizational objectives, goal setting and tracking can be applied to individual and team performance within a CRM system. For example, setting goals for sales representatives based on metrics like sales targets, conversion rates, and average deal size can help monitor their performance and drive productivity. By tracking these metrics, management can provide targeted coaching and support to help individuals and teams reach their goals. Regularly reviewing progress and providing feedback are integral to this process, ensuring everyone is on track and taking the necessary steps to succeed. CRM implementation can yield significant results and drive overall organizational success by aligning individual and team goals with business objectives.

Operational efficiency is another critical aspect of CRM success. Setting goals to improve efficiency can streamline processes and reduce costs. By tracking metrics related to productivity, such as order fulfillment time, customer service response time, and inventory turnover, management can assess the efficiency of their operations. If specific metrics do not meet the desired targets, it may indicate bottlenecks or areas that require process improvements. Using these metrics to guide decision-making, management can make data-driven adjustments and optimize operations to achieve efficiency goals. By continuously monitoring and optimizing key operational metrics, administrators and project teams can drive efficiency improvements and enhance overall CRM system performance.

Goal setting and tracking with KPIs and metrics are key for measuring the success of a CRM system. Leaders, project teams, and administrators involved in CRM implementation must take advantage of the power of these tools to set clear and quantifiable goals, monitor progress, and adjust strategies accordingly. By focusing on revenue growth, customer retention, employee performance, and operational efficiency, businesses can drive their CRM initiatives forward and achieve tangible results.

Transparency And Accountability; Transparency and accountability are crucial to measuring success throughout implementing and operating a CRM system. Intrinsic to this success is presenting and communicating Key Performance Indicators (KPIs) and metrics transparently within the CRM. Transparent communication of metrics and KPIs allows stakeholders to understand their roles and responsibilities in achieving desired outcomes, fosters a culture of accountability, and encourages data-driven decision-making. Measuring success with tangible data can be applied to various organizational areas, including sales, customer service, project management, and financial performance.

In a sales department, presenting KPIs and metrics such as sales targets, conversion rates, and revenue generated within the CRM can significantly improve team performance. This practice sets clear expectations for sales representatives and enables them to track their progress in real-time. By visualizing their performance, salespeople can take ownership of their targets, identify areas for improvement, and make data-driven decisions to drive better results. It also allows management to evaluate individual performance, provide constructive feedback, and recognize top performers. With this level of transparency, everyone can see the big picture and work collaboratively towards achieving shared goals. In short, transparently presenting sales metrics fosters a culture of accountability and encourages the entire team to strive for shared success.

Presenting customer service KPIs and metrics like response time, customer satisfaction scores, and first-call resolution rates transparently within the CRM is crucial for fostering accountability and continuous improvement. When these metrics are openly shared within the CRM, customer service representatives understand the expected standards and can strive to meet and exceed them. Transparent communication also enables managers to identify bottlenecks or training needs, provide targeted coaching, and recognize employees who consistently deliver excellent service. Setting transparent customer service performance metrics catalyzes driving an internal performance culture and fosters an environment of individual accountability for customer excellence.

Setting, tracking, and presenting project management metrics like project completion rates, milestones achieved, and resource utilization transparently within the CRM promotes transparency and accountability. By making this information visible to all stakeholders, project teams can understand the project's progress, identify areas of concern, and take appropriate actions to keep projects on track. Transparent communication of project metrics also helps stakeholders evaluate project health, allocate resources effectively, identify potential risks, and make informed decisions. When everyone involved in a project knows how their performance contributes to overall success, a strong sense of accountability is fostered, leading to better project outcomes. In short, transparency in project management metrics is critical to measuring success, maintaining solid relationships, and fostering a culture of collaboration and accountability.

Transparently presenting financial KPIs and metrics within the CRM, such as revenue, profit margins, and cash flow, ensures stakeholders have a clear view of the organization's financial health. This transparency promotes accountability by providing decision-makers know their responsibility in driving financial performance. Regular review of these metrics by management can lead to strategic discussions, identifying areas for cost optimization, making informed investment decisions, and holding all relevant stakeholders accountable for contributing to financial success. By utilizing transparent financial metrics within the CRM, businesses can foster a culture of accountability for financial success and empower decision-makers to drive their businesses forward confidently.

Transparency and accountability are not simply good business practices; when measuring a CRM, they are essential components of the system's success, they are essential components. Leaders, project teams, and administrators involved in CRM implementation must use transparent performance metrics to set clear and quantifiable goals, monitor progress, and adjust strategies accordingly. Highlighting potential pitfalls and the importance of due diligence in a cautionary tone is vital to a successful CRM system. By holding everyone accountable for their performance, teams can foster a culture of continuous improvement, leading to enhanced customer satisfaction, deeper insights, and better decision-making. A CRM system supported by transparent and accountable performance metrics is poised to enhance overall organizational success.

Common KPIs And Metrics For CRMs; Some of the most common KPIs and metrics used to measure the effectiveness of a CRM include:

Sales revenue: *Total revenue generated from sales activi-*

ties, often broken down by product, region, or sales team.

Conversion rate: *The percentage of leads converted into customers, indicating the effectiveness of sales and marketing efforts.*

Customer acquisition cost (CAC): *The total cost of acquiring a new customer, including marketing, sales, and onboarding expenses.*

Customer lifetime value (CLV): *The projected revenue a customer will generate over their relationship with the business.*

Customer retention rate: *The percentage of customers retained over a specific period, indicating the effectiveness of customer service and relationship management efforts.*

Average response time: *The average time it takes for customer queries or issues to be addressed, reflecting the efficiency of customer service processes.*

Customer satisfaction score (CSAT): *A measure of customer satisfaction based on feedback and ratings, indicating the overall quality of customer experience the business provides.*

Challenges In Identifying And Presenting KPIs And Metrics; Measuring the success of a CRM system requires careful consideration and planning in selecting and presenting KPIs and metrics. However, this process has challenges, and leaders, project teams, and administrators must navigate these obstacles effectively. Let's delve into the critical challenges of identifying and presenting KPIs and metrics within a CRM system.

- **Selecting Relevant KPIs and Metrics:** One of the primary challenges lies in selecting the most relevant and meaningful KPIs and metrics that align with the business's goals and objectives. With a vast array of potential metrics, figuring out which ones will provide actionable insights and drive decision-making can be challenging. It is critical to review the organization's specific requirements, ensuring that the selected metrics are aligned with strategic priorities. Conducting thorough research, gathering input from key stakeholders, and analyzing industry best practices can aid in the selection process.

- **Data Quality and Accuracy**: The effectiveness of KPIs and metrics within a CRM system heavily relies on the quality and accuracy of the underlying data. Poor data quality can lead to misleading or inaccurate insights, undermining the value of these metrics. Incorrect or incomplete data can arise from various sources, such as human error during data entry, outdated information, or data integration issues. Organizations must invest in robust data governance practices, including data cleaning, validation, and integrity. Implementing data quality checks and regular data audits can help identify and rectify any issues promptly, ensuring the reliability and credibility of the KPIs and metrics presented within the CRM.
- **Integration with Other Systems:** Another significant challenge lies in integrating KPIs and metrics from multiple systems and data sources to provide a comprehensive view of performance. Organizations may have data residing in various systems such as ERP, financial management software, or customer survey tools, making it challenging to gather and present a holistic view of performance metrics within the CRM. Successful integration requires a thoughtful and well-executed data integration strategy that ensures data flows seamlessly between systems. Leveraging integration tools, such as APIs or middleware, can help streamline data integration processes and provide accurate and timely updates of KPIs and metrics within the CRM. Project teams and administrators must collaborate with IT experts to understand the dependencies and complexities of integrating disparate systems.

- **Visual Presentation and Communication:** Effectively presenting KPIs and metrics within a CRM clearly and appealingly can prove challenging, mainly when dealing with complex data sets and multiple stakeholders with varying levels of data literacy. The goal is to transform raw data into meaningful, digestible, easily understandable, and actionable information. Representations such as charts, graphs, and dashboards can enhance data comprehension and enable stakeholders to make informed decisions. However, it is essential to balance visual appeal and data accuracy. Presenting too much information can overwhelm users, while overly simplistic visuals may lack the necessary depth of analysis. Strive for clarity, simplicity, and relevance in your visual presentations, tailoring them to different stakeholders' specific needs and preferences.

Identifying and presenting KPIs and metrics within a CRM system pose several challenges that must be addressed to measure success effectively. This journey requires a careful selection process to identify the most relevant metrics, a focus on data quality and accuracy to ensure trustworthy insights, an integration strategy to consolidate data from multiple systems, and a thoughtful visual presentation to communicate effectively. Leaders, project teams, and administrators should approach this process diligently, seeking expert guidance when necessary and making informed decisions to achieve optimal outcomes. By overcoming these challenges, organizations can leverage the full potential of their CRM system and truly measure success in a meaningful and impactful way.

By addressing these challenges and effectively presenting and communicating KPIs and metrics within a CRM, businesses can make informed decisions, evaluate performance, allocate resources effectively, and ultimately drive the success of their CRM initiatives.

2

Comfort

Comfortability is subjective; what is convenient for one person with other skills and behaviors can be inconvenient for another.

CRM systems have revolutionized how businesses manage customer and prospect relationships (Accounts, Opportunities, and Leads). By creating a vast pool of frequently previously non-existent informational structures, CRMs enable organizations to access valuable insights and make data-driven decisions that ultimately drive success. However, achieving this success relies heavily on users' adoption and acceptance of the CRM. A quick user acceptance rate is crucial to create the ongoing source of inputs necessary to make such a pool of information.

The key to aligning adoption and acceptance with a user's success lies in demonstrating how their use of the CRM benefits the entire organization and directly impacts their job performance and responsibilities. It is essential to highlight that by contributing to and utilizing the CRM, users effectively create a circular pattern that rewards their efforts with better information, allowing them to excel in their roles.

One of the primary ways in which CRMs benefit users is by streamlining and automating various tasks, freeing up valuable time and resources. By inputting data and using the system to manage customer interactions, users can reduce the time spent on manual tasks and instead focus on more strategic and revenue-generating activities. This increased efficiency can improve job performance, satisfaction, and tremendous organizational success.

Moreover, CRMs give users access to a wealth of information to help them make informed decisions. Users can gain valuable insights that guide their actions and strategies by inputting data and using the system to analyze customer behavior, preferences, and trends. For example, a sales representative can use the CRM to identify potential upsell opportunities based on a customer's purchase history, or a marketing professional can use the system to segment customers for targeted campaigns. By leveraging this information, users can make more informed decisions and achieve better results in their roles.

In addition to providing insights and streamlining tasks, CRMs facilitate collaboration and communication among team members. By creating a centralized repository of customer data, the CRM ensures that all users have access to the same information, which can help prevent communication breakdowns and ensure that everyone is working from the same playbook. This increased collaboration can lead to more effective teamwork and ultimately contribute to the success of both individual users and the organization.

To foster the adoption and acceptance of CRM, it is essential to provide users with the necessary training and support to ensure they can effectively use the system to its full potential. Organizations can empower users to take full advantage of the CRM and its capabilities by offering comprehensive training programs, ongoing support, and resources. This, in turn, will contribute to the circular pattern of success, as users proficient in using the system will be more likely to input data and utilize the system to make better decisions.

Another crucial aspect of promoting the adoption and acceptance of CRM is to ensure that it is user-friendly and customized to meet the needs of individual users. By designing a CRM that is intuitive, easy to navigate, and modified to meet the specific requirements of each user, organizations can encourage users to embrace the system and incorporate it into their daily workflows. This will further contribute to the circular pattern of success, as users who find the system easy to use and relevant to their roles will be more likely to input data and use the system to drive their success.

Recognizing and rewarding users who actively contribute to and utilize the CRM is also important. By acknowledging the efforts of users who input data, share insights, and leverage the system to achieve better results, organizations can create a culture of success that encourages others to follow suit. This can be achieved through various means, such as offering incentives, providing recognition, or implementing gamification strategies to motivate users to engage with the CRM.

Aligning the adoption and acceptance of a CRM with users' success is a critical factor in realizing the system's full potential and driving organizational success. By highlighting the circular pattern of success that rewards users for their contributions and utilization of the system, organizations can create a culture of engagement and collaboration that ultimately benefits both individual users and the organization. By streamlining tasks, providing valuable insights, fostering cooperation, offering training and support, and recognizing users' efforts, organizations can ensure that their CRM becomes an indispensable tool for driving success and achieving better results.

3

Impact Of Unresolved Resistance On CRM Success

If employee resistance to the CRM system is not effectively addressed, it can have a significant impact on the success of the implementation:

Low User Adoption; unresolved resistance can result in low user adoption of the CRM system. Employees may be reluctant to use the CRM or may use it ineffectively, resulting in the system being underutilized. This can limit the potential benefits of the CRM and reduce the return on investment.

Inefficient Processes; resistance to the CRM system can lead employees to continue using their existing techniques and tools. This can result in inefficiencies and duplication of effort, as the organization cannot fully leverage the capabilities of the CRM. Inefficient processes can hinder productivity and prevent the organization from maximizing the benefits of the CRM system.

Reduced Data Quality; resistance to the CRM system can also impact the quality of data entered into the system. Employees may input inaccurate or incomplete data, resulting in reduced data quality. This can undermine the reliability of insights and analytics generated from the CRM, making it difficult for the organization to make informed business decisions.

Decreased customer satisfaction: if employees are not fully utilizing the CRM system, they may struggle to provide the level of service and support that customers expect. This can lead to decreased customer satisfaction and potential loss of business. The CRM system is designed to enhance customer relationships and improve the overall customer experience. Therefore, it is essential to address resistance to ensure that employees can leverage the CRM effectively in their interactions with customers.

Addressing employee resistance to CRM implementation is vital for the project's success. Organizations can overcome resistance by adopting strategies such as identifying difficult leaders, communicating the benefits of CRM, involving employees in the implementation process, providing comprehensive training, addressing job security concerns, and introducing the CRM gradually. Failure to address employee resistance can significantly impact CRM success, including low user adoption, inefficient processes, reduced data quality, and decreased customer satisfaction. By proactively addressing opposition, organizations can maximize the benefits of CRM and drive positive change within the organization.

Understanding the reasons behind employee resistance to CRMs and adopting strategies to address these concerns is crucial for ensuring the success of the implementation. By effectively managing resistance, organizations can maximize user adoption, improve efficiency, and ultimately achieve the desired benefits of CRM.

4

Fear Of Excessive Productivity Monitoring

Warning: If, in fact, you are monitoring people's performance - it's essential to approach it carefully and transparently. Drawing from my wealth of experience, I have learned that altering tactics and communication strategies can be necessary to ensure success. However, it's crucial to tread lightly and maintain honesty throughout the process.

IN THE REALM OF CRM, and any organizational change initiative, monitoring productivity is essential to gauge the implementation's effectiveness and identify improvement areas. By adjusting your tactics and communication strategy, you can better analyze and track the productivity levels of individuals involved in the implementation.

However, it's crucial to proceed with caution when monitoring productivity. People may feel uneasy or resistant if they sense their every move is being watched or scrutinized. As a mentor in this field, I strongly advise a balanced approach that respects privacy while fostering accountability.

Transparency is critical when it comes to monitoring productivity. Openly communicate with your team members about the importance of data-driven decision-making and the need to understand the impact of CRM implementation on productivity. Be clear about what data will be collected, how it will be used, and the ultimate goal of improving efficiency and effectiveness.

While tracking productivity and gathering relevant data is essential, it's equally important to prioritize ethical practices and respect for individuals' privacy. Strive to balance monitoring productivity and maintaining a culture of trust and collaboration.

When communicating with your team members about productivity, always tell the truth. Honesty is crucial for building trust within the organization. However, remember that certain confidential or sensitive information must be handled carefully. You can be transparent about the CRM implementation's goals, progress, and impact without compromising sensitive details.

Remember, the success of a CRM implementation goes beyond technicalities – it depends on the people involved and their commitment to the process. By carefully adjusting your tactics and communication strategies, monitoring productivity respectfully and transparently, and ensuring honesty to the greatest extent possible, you can create an environment that motivates and engages your team members, leading to a successful CRM implementation that drives efficiency and cultivates strong customer relationships.

Another concern that can breed resentment is the fear of the CRM monitoring their productivity too closely. Users may worry that every action they take, every call they make, and every minute they spend on a task will be meticulously tracked and analyzed. This fear may stem from a lack of understanding that CRMs aim to improve efficiency and provide valuable insights rather than continuously monitor and micromanage individual performance.

Fear of over-monitoring of productivity by a CRM is a common concern many users have. Still, with the right approach, we can prove the need for the system, improve buy-in, and navigate potential pitfalls.

PROVING THE NEED; To address the fear of over-monitoring, we must establish the need for a CRM. Emphasize that the system isn't designed to micromanage individual performance but to improve efficiency, streamline workflows, and provide valuable insights. Make a case for how the system can help users prioritize tasks, automate repetitive tasks, and save time for more critical work. Provide examples of how the system has helped other organizations optimize their operations and drive positive outcomes. Explain how the information gathered through the CRM can help managers make data-driven decisions and improve overall performance.

IMPROVING BUY-IN; To enhance user buy-in, we must communicate effectively and foster a sense of trust. Address the fear of over-monitoring head-on and explain the system's intended purpose. Be transparent about how data is collected, who has access to it, and how it will be used. Provide training and support to ensure users are comfortable with the system and understand how it can benefit them. Consider appointing ambassadors to advocate for the design and champion their positive experiences with their peers. Encourage user feedback and input to shape the system according to their needs.

POTENTIAL PITFALLS; When it comes to potential pitfalls, we need to take a proactive approach and have a plan to mitigate risks. One potential pitfall is that users may feel overwhelmed by the amount of data the system collects. To address this, provide clear and concise reports highlighting the most critical information and making it easily accessible. Additionally, establish privacy policies and permissions to ensure that sensitive data isn't shared without proper authorization. Another potential pitfall is a lack of adoption among users. This can occur if they don't see immediate benefits or feel overwhelmed by the learning curve. To avoid this, provide ongoing training and support, and encourage active participation by involving users in the selection and implementation process.

5

Perception Of Stalking

Customers and Employees Value Their Privacy

In today's fast-paced business landscape, it's more important than ever for businesses to prioritize customer and employee concerns regarding privacy. Especially when implementing CRM solutions, we need to be extra careful not to infringe on customers' privacy and potentially alienate them.

CRMs offer various valuable tools that can help enhance customer engagement and loyalty. Still, we need to ensure we're not overwhelming our customers or causing them to feel like their precious privacy is being invaded. Proactively demonstrating respect for customer privacy is critical to building and maintaining trust, mainly when dealing with external customers.

But let me tell you; it doesn't stop there. As we delve further into this topic, you'll see how internal customers, such as salespeople, are also highly affected by the trust users in the CRM's ability to respect their protected client information. This is especially true of those who have worked in the industry for years and amassed a network of 30 years of contacts. To them, trust in the system's ability to handle confidential information is the foundation of their business relationships.

As businesses continue to adopt and rely on CRM solutions, keeping privacy concerns in mind is essential. We must demonstrate respect for customer privacy in a proactive way that builds and maintains trust. You can create CRMs that work for external customers and internal team members by incorporating all the necessary privacy measures.

It's all about finding that perfect balance between enhancing customer experience through CRMs and maintaining trust and privacy. Take a step back, listen to your customers and employees, and ensure you implement a CRM solution that keeps privacy concerns at the forefront while providing excellent customer service and loyalty.

Here are key ways to help you build trust in your system:

Be Clear About What Data Is Collected And Why; It's essential to bring transparency into your data collection processes. Ensure your customers know exactly what data is being collected and why. By being upfront and transparent with your customers, you'll be able to establish trust with them - which is one of the most critical aspects of the modern business landscape.

Provide Customers With An Opt-Out Option; that's easy to find, use, and understand. People love options, and it's no different when it comes to their data. Make sure you provide your users with an easy-to-find opt-out option. By doing so, you are offering them control over their data and creating trust.

Ensure All Users Proactively Know Your Service's Privacy Policy; (and periodically after that). Educate your users and ensure they are aware of the privacy policy. When they sign up for your service, highlight and emphasize how you handle their data.

Ensure Your Privacy Policy Is Easy To Find, Use, And Understand; A confusing privacy policy is more harmful than not having one at all. Ensure your privacy policy is written in plain language and avoids technical jargon. Use simple and easy-to-understand terms to make it easy for everyone, novice or expert, to comprehend.

Ensure You Have A Clear And Consistent Process For Handling Data Security Breaches_(and ensure this is communicated). No system is invincible, but ensuring that you have a clear and consistent process for managing data security breaches is essential. Ensure you communicate this process internally and, if necessary, externally.

Wonky Belief Debunking

Warning: This entire section is an alarm. It is essential to understand and proactively address and remedy.

ONE OBVIOUS BUT common common misconception is the belief that customers desire increasing more contact and close monitoring from the companies they patronize. While occasional incentives, personalized offers, and attempts at building long-term relationships can be effective strategies for customer retention, inundating customers with excessive offers and communication can have the opposite effect, causing frustration and dissatisfaction.

To delve deeper into this issue, one must consider the diverse factors contributing to external customer happiness. For instance, if one were to inquire about brand loyalty initiatives that could make customers happier, the responses would likely be varied, depending on the type of business and the specific requirements of individual customers. While an eCommerce intermediary might receive suggestions for faster delivery and enticing discounts, a luxury jewelry store could expect to hear ideas about dedicated customer service and exclusive in-store or digital experiences.

These customer preferences and expectations can differ significantly between industries and companies within a particular industry and individual customers within a specific business. This complexity highlights the need for organizations to recognize that CRM solutions and strategies that might have yielded remarkable results for one company may not necessarily be suitable or effective for another.

The crux of customer-company relationships lies in their mutually beneficial nature. A company may strive to foster stronger bonds with its loyal customer base. Still, it is essential to consider whether those customers reciprocate the desire to establish and nurture relationships with the organization. Establishing ties with customers who trust the company and actively engage with its brand. Failing to do so could potentially result in losing them to competitors.

Conversely, attempting to forge relationships with disinterested or skeptical customers can have negative consequences, as the company's actions may be perceived as intrusive or unwarranted. In such cases, prospective customers may become irritated and become vocal critics, significantly damaging the organization's brand reputation.

Numerous examples exist of customers who have recounted invasive or uncomfortable experiences with overzealous CRM practices. The key takeaway from such anecdotes is that the mere ability of managers to reach customers does not automatically warrant doing so. The nature of the customer's relationship with the brand, not the CRM software's capabilities, should dictate the company's approach to engagement.

Demonstrating respect for privacy when implementing CRM solutions is of paramount importance. Organizations must consider the needs and preferences of external customers and the potential pitfalls of indiscriminate or overbearing CRM practices. By purposefully and proactively presenting themselves as respectful of customer privacy and tailoring CRM strategies accordingly, businesses can foster trust, enhance customer satisfaction, and build lasting, mutually beneficial relationships with their clientele, ensuring long-term success and growth in an increasingly competitive marketplace.

6

Unwillingness To Make Their Work More "Transparent"

One of the most significant yet frequently overlooked obstacles to successful CRM implementation lies in the spheres of human nature and workplace dynamics. Interestingly, it's not uncommon for sales personnel to harbor feelings of skepticism or resistance towards CRMs. They often perceive shared digital solutions as intrusive entities potentially inclined to usurp their hard-earned client relationships and intricate sales techniques. To tackle this, fostering an environment where CRM solutions are perceived as supportive business partners rather than intrusive platforms is quintessential.

In the realm of CRM, a considerable amount of information needs to be captured regularly. This includes but is not limited to:

- Contact Details,
- Communication Timeline,
- Task Planning, Execution, and Outcomes
- Intellectual Property.

A CRM IS DESIGNED to meticulously track the existence, quantity, and quality, thereby flagging any inconsistencies or gaps. This feature safeguards against data discrepancies, improving client information's accuracy and reliability.

However, it is the very functionality of this system that serves as a double-edged sword. While the management team benefits from this insightful and transparent data pool, enabling them to address several key organizational aspects, such as refining CRM processes, safeguarding against loss of customer information in the event of an employee transition or resignation, and assessing employees' performance objectively and identifying those not meeting expectations.

Certain employees, especially those in customer-facing roles, might perceive this as a threat. This apprehension often manifests as resistance towards implementation or optimal utilization of CRMs.

Why does this phenomenon occur in any case? The crux of the matter lies in the inherent nature of human behavior and workplace dynamics. Roles like sales and customer service are complex and involve highly diverse and unpredictable human behaviors. The CRM's tracking and monitoring capabilities make their performance more tangible, transparent, and thus measurable. This objective evaluation leaves little room for justifying underperformance or failures on subjective grounds like demanding clients or unpredictable market trends, leading to discomfort amongst such employees.

THERE ARE THREE PRIMARY, albeit contrasting, reasons that underpin the resistance to transparent performance assessments:

- Irreplaceability
- Perceived Productivity Monitoring
- Irreplaceability

MOST CEO'S can recall a scenario where a key account manager or a critical team leader in sales, marketing, or service held the company hostage with a request, subtly implying, "Accept my demands or risk losing me, along with my invaluable client relationships." The predicament extends beyond the loss of experienced personnel — the potential departure of significant customer data.

An efficient team leader often holds a lot of customer information ranging from basic contact details to intricate data on client needs, desires, and concerns. If such an individual were to resign or shift to a competitor, the odds of replenishing this personalized customer data without their cooperation are slim, potentially hindering the organization.

Simultaneously, these individuals could use the same information to boost their value at a competing firm. Hence, this creates a situation where the senior management is caught in a tussle between succumbing to the employee's ultimatum or risking a significant business setback.

For this reason, sales, marketing, and customer service managers often feel compelled to limit disclosures of detailed customer data, even when such data accumulation is an organic result of their job role. They perceive customer information as a means to enhance their market value.

Given these factors, it's understandable why certain employees view CRM solutions, notorious for their transparent and comprehensive data management features, as severe threats to their job security and approach CRMs with trepidation or outright resistance.

Therefore, to promote more enthusiastic adoption of CRMs, organizations must ensure that their sales teams perceive CRMs as trusted business partners that aid, rather than undermine, their endeavors. Efforts must be dedicated to dispelling fears or misconceptions surrounding privacy or job security concerns associated with CRMs, ensuring that CRM becomes a strategic ally fostering customer relationships and driving sales success. Emphasizing these systems' mutual benefits, including enhanced efficiency, customer satisfaction, and sales performance, will promote their acceptance as valuable tools for business growth rather than perceived threats to job security.

Irreplaceability

Unfortunately, the only way to combat "irreplaceability" is by administrative interventions, and only if top-level executives are adamant about it. If you do not have such assistance initially, the CRM investments will be jeopardized, and the CRM project manager may "feel" the failure in many ways.

WHEN CONSIDERING ways to gain your system's and processes' trust, it's essential to carefully evaluate the pros and cons of implementing a CRM. One of the key benefits of leveraging CRM is improved business development activities via supercharged client relationship management insight. CRMs provide a centralized platform for storing and managing all customer-related data, which can significantly assist in better CRM.

However, it's crucial to address employees' concerns about the potential misuse of sensitive data stored within the system. To alleviate these concerns, **publicly and loudly emphasize** the stringent security protocols inherent in CRMs that safeguard against unauthorized data access or misuse. By highlighting the protective measures, you can help alleviate any fears or reservations your team may have.

Another aspect to consider is the use of CRM for effective task management. CRMs offer task management and synchronization features that can help streamline workflow and increase productivity. However, some employees may perceive this as a mechanism for increased surveillance of their workflow. To address these concerns, **highlight the added efficiency and convenience** these features can bring to their daily work routines. By framing it as a tool to support their efforts, you can help overcome any resistance or apprehension.

It's also important to explain the sales forecasting power of CRM. CRMs can **analyze past sales data to provide accurate sales forecasts, enabling better sales strategies**. However, some employees may be apprehensive about automated predictions replacing their judgment. To alleviate these concerns, reinforce that these predictions are meant to **supplement, not replace, their market insights and experience**. By emphasizing the importance of human judgment and the role of CRM as a support tool, you can gain their trust and buy-in.

One of the unique capabilities of CRM is its utility in personalizing client interactions. The **detailed client profiles** stored in CRM can enable highly personalized interactions, fostering stronger relationships. However, it's essential to address concerns about privacy and potential invasiveness. Assure your team that privacy norms will be strictly adhered to while handling customer data, emphasizing the importance of respect and professionalism when personalizing interactions.

CRMs also play a crucial role in **facilitating collaboration and teamwork**, particularly in harnessing the potential of junior team members. CRM enables collaboration, allowing junior team members to share the workload effectively, ultimately leading to more business. However, some employees might fear involving junior members will dilute their client relationships. To address these concerns, ensure clear communication about role definitions and boundaries, emphasizing each team member's value. You can alleviate any resistance or hesitation by fostering a collaborative environment focused on growth and success.

Another advantage of CRM is its ability to **reduce administrative work** through automation. CRMs can automate numerous administrative tasks, freeing up staff for more critical tasks. However, the team might fear job redundancy due to automation. To counter this perception, point out that automation is meant to eliminate mundane tasks, enabling employees to concentrate on more meaningful, value-added work. By emphasizing the opportunity for growth and development, you can help them see automation as a tool to enhance their skills rather than a threat to their job security.

CRM also serves as a powerful tool for professional development. It provides valuable insight into individual performance and areas of improvement, aiding in personal development. However, some employees may view it as a surveillance tool to monitor their performance. To address these concerns, reinforce the focus on improvement and growth rather than surveillance or penalization. By highlighting the potential for personal and professional growth, you can help employees embrace CRM as a tool for development rather than management.

Transparency and trust are crucial elements of successful CRM implementation. CRM's centralized data storage promotes transparency and trust among team members. However, some staff members may feel uncomfortable with the transparency of their client relationships. To alleviate these concerns, highlight the benefits of shared knowledge, trust, and collaboration that arise from transparency. Emphasize how this transparency can lead to more effective teamwork and better client outcomes.

CRMs also play a vital role in client retention. With their analytical capabilities, CRMs can help **identify at-risk client relationships**, enabling preemptive actions to improve client retention. However, staff might feel that their hard-earned relationships are undervalued if a system can identify at-risk clients. To address this, reiterate that CRM is a preventative tool, supplementing their relationship-building efforts. By highlighting the role of CRM as a support mechanism rather than a replacement for a personal touch, you can help your team see the value it brings to client retention efforts.

Lastly, CRM serves as a continuity safeguard. It **safeguards against losing client information during personnel changes, ensuring continuity in client relationships**. However, some staff may consider this a ploy to make them easily replaceable. To dispel this perception, emphasize the overall business need for continuity and stress that the tool is not designed to undermine its value or uniqueness. By framing CRM as a tool to support the team and provide seamless transitions for clients, you can overcome resistance and gain their trust.

In summary, gaining the trust of your team in implementing CRM requires addressing their concerns and highlighting the benefits that CRM brings to various aspects of their work. You can guide your team through potential challenges and ensure a successful CRM implementation by emphasizing security protocols, efficiency gains, and growth opportunities.

7

Post-Implementation CRM Maintenance And Optimization

GUIDES, PROCESS, PLANS & TRAINING

A CRM is pivotal in business success as it efficiently manages customer interactions, sales pipelines, and marketing campaigns. However, a CRM demands regular maintenance and optimization for years to ensure effectiveness and efficiency. This is where a maintenance and optimization guide comes into play. In this witty and professional exploration of CRM care, I will delve into the value of crafting a comprehensive guide that will keep your CRM in excellent condition, ensuring its longevity and continued success.

The importance of a maintenance and optimization guide cannot be overstated. It is the foundation for your CRM's long-term success. A well-crafted guide can help you identify and remedy potential issues before they become significant problems, saving you time, money, and frustration in the long run. You can minimize downtime, reduce costs, and maintain a seamless customer experience by staying ahead of the curve.

Moreover, a maintenance and optimization guide can help you maximize the return on your CRM investment. Regularly reviewing and optimizing your CRM ensures you get the most value, streamlining processes, and maximizing efficiency. This will likely lead to increased revenue, happier customers, and a more successful business overall.

Staying agile and adaptable is crucial in today's rapidly changing business landscape. A maintenance and optimization guide can help you respond to shifts in customer preferences, market trends, and technological advancements, ensuring your CRM remains relevant and practical. By regularly updating and refining your CRM, you can stay ahead of your competition, thus allowing you to continue to deliver exceptional customer experiences.

Furthermore, a maintenance and optimization guide can help instill a continuous improvement mindset within your organization. Regularly reviewing and refining your CRM can encourage a culture of innovation, growth, and customer-centricity. This can increase employee productivity, improved collaboration, and a more successful business overall.

So, what should be included in a maintenance and optimization guide to ensure a healthy CRM? First and foremost, it's essential to conduct periodic audits of your CRM to identify any discrepancies, inefficiencies, or areas for improvement. This could include reviewing data quality, user adoption, system performance, and integrations with other software. By regularly assessing the state of your CRM, you can ensure it remains in peak condition and continues to deliver value to your organization.

In addition to regular audits, it's crucial to establish clear guidelines for data entry, storage, and management to ensure the accuracy and consistency of information within your CRM. This could include creating standardized data entry protocols, implementing data validation rules, and regularly cleansing and deduplicating records. By maintaining a high level of data quality, you can ensure your CRM remains a reliable and trustworthy source of information for your organization.

Ongoing user training and support are critical components of a successful maintenance and optimization guide. By providing your team with the necessary resources and assistance, you can ensure they are well-equipped to use the CRM effectively and efficiently.

This leads to increased user adoption, improved collaboration, and a more successful business overall.

Process optimization is another critical aspect of maintaining a healthy CRM. You can continuously review and refine your CRM processes to ensure they remain efficient and aligned with your business objectives. This could involve streamlining workflows, automating manual tasks, and eliminating bottlenecks. By regularly updating and improving your CRM processes, you can ensure your system remains agile and adaptable, ready to respond to the ever-changing needs of your organization.

Staying up-to-date with the latest CRM updates and upgrades is essential for maintaining a healthy CRM. Regularly reviewing release notes, participating in beta testing, and scheduling timely system updates can ensure your organization is leveraging the most advanced features and capabilities. This, in turn, can lead to improved system performance, increased user adoption, and a more successful business overall.

Performance measurement is another critical component of a successful maintenance and optimization guide. Establishing clear metrics and benchmarks allows you to measure your CRM's success and identify improvement areas. This could include tracking quantitative factors, such as revenue growth and cost savings, and qualitative factors, such as customer satisfaction and employee engagement. By regularly assessing the performance of your CRM, you can ensure it continues to deliver value to your organization and drive business success.

Finally, developing a robust change management strategy ensures a smooth transition when implementing new features, processes, or integrations within your CRM. This could include communicating the benefits of the changes, providing training and support, and monitoring the impact on system performance and user adoption. By effectively managing change, you can ensure your CRM remains agile and adaptable, ready to respond to the ever-changing needs of your organization.

The value of crafting a comprehensive maintenance and optimization guide for your CRM cannot be overstated. Investing time and effort into creating a clear roadmap for maintaining your CRM's health and performance can ensure it remains a valuable asset for years. So, approach this journey with wit and professionalism, and enjoy the benefits of a well-maintained and optimized CRM for years to come.

As we navigate the ever-evolving landscape of business technology and customer preferences, the importance of a well-maintained CRM becomes increasingly apparent. A CRM is much more than a mere tool; its existence should insinuate the overhaul of business processes and become the lifeblood of a customer-centric organization. By creating a maintenance and optimization guide that addresses the various aspects of CRM health, you invest in your business's future success and longevity.

Ultimately, creating a maintenance and optimization guide is a testament to your organization's commitment to excellence, innovation, and customer satisfaction. It reflects your dedication to providing the best possible customer and team experience. By crafting a comprehensive guide, you are setting the stage for a CRM that not only stands the test of time but continues to drive your business forward in a world of ever-changing customer expectations and technological advancements.

So, as you create a maintenance and optimization guide for your CRM, approach the task with wit, professionalism, and a keen understanding of the value it holds for your organization. Doing so will lay the foundation for a healthy, efficient, and effective CRM for years, ensuring your business's continued success and growth.

A comprehensive CRM maintenance and optimization guide is crucial for several reasons:

Ensuring data accuracy; A well-maintained CRM ensures that the data stored within it is accurate, up-to-date, and free from duplicates. A quality guide will provide detailed instructions on regular data cleansing and deduplication tasks, which are essential for maintaining data integrity and enabling informed decision-making.

Optimizing system performance: Regular maintenance and optimization tasks help to keep the CRM running smoothly, with minimal downtime or performance issues. A quality guide will outline the steps for monitoring system performance, identifying potential bottlenecks, and implementing optimizations to improve response times and overall user experience.

Maintaining data security and compliance; Data security and compliance are critical concerns for any organization, particularly those handling sensitive customer information. A comprehensive CRM maintenance and optimization guide will include instructions for ensuring data security, such as regularly updating user permissions, monitoring potential breaches, and adhering to relevant data protection regulations.

Streamlining workflows and processes; A well-maintained CRM enables organizations to simplify their workflows and processes, increasing efficiency and productivity. A quality guide will guide reviewing and updating workflows, automating repetitive tasks, and ensuring that the CRM remains aligned with the organization's evolving needs and objectives.

Maximizing the value of the CRM investment; By following a comprehensive maintenance and optimization guide, organizations can maximize the value of their CRM investment by ensuring that the system remains in optimal condition and continues to deliver the desired results. This includes integrating other systems, optimizing reporting capabilities, and proactively addressing issues.

CAUTION - Some potential problems exist if a quality guide is not created and followed. Don't skimp on this effort, and don't let others walk away from the project before completion.

IF A QUALITY CRM maintenance and optimization guide is not created and followed. In that case, several problems may arise, including:

Data Degradation; Without proper data management, the CRM may become cluttered with duplicate or outdated information, leading to inefficient processes, poor decision-making, and reduced customer satisfaction. This can ultimately result in lost sales opportunities and negatively impact the organization's bottom line.

System Performance Issues: Failure to monitor and optimize system performance can result in slow response times, crashes, and poor user experience. This can lead to decreased productivity, user frustration, and potential loss of customers due to dissatisfaction with the CRM.

Security Breaches; Neglecting data security and compliance can expose the organization to potential data breaches, resulting in legal and financial consequences and damage to the organization's reputation. In the worst-case scenario, a security breach could lead to the loss of sensitive customer data, with severe implications for the organization and its customers.

Inefficient Workflows; Without regular review and updates to workflows and processes, the CRM may become outdated and inefficient, hindering productivity and customer satisfaction. This can increase costs, as employees spend more time on manual tasks and less on value-added activities, such as nurturing customer relationships or closing sales deals.

Integration Failures; Failing to maintain integration with other systems can lead to data silos and disjointed processes, negatively impacting the overall effectiveness of the CRM. This can result in reduced visibility into customer data and interactions, making it more difficult for organizations to deliver personalized and targeted marketing campaigns or provide exceptional customer service.

A COMPREHENSIVE CRM maintenance and optimization guide ensures a CRM's smooth operation and ongoing success. By following a well-structured directory, organizations can proactively address potential issues, maintain system performance, and maximize the value of their CRM investment. Failure to create and follow a quality guide can result in various problems, from data degradation to system performance issues, ultimately impacting the organization's ability to manage customer relationships and drive growth effectively.

8

Reluctance To Change And Adapt

Users, especially experienced and older professionals, may resent the CRM simply because it represents a significant change in how they work. They may be accustomed to their old processes and resistant to adopting new technology. This reluctance to change and adapt can lead to resentment towards the system, as it disrupts their established routines and requires time and effort to learn and implement. Many organizations face This very common hurdle, so don't feel too surprised if it slaps you in the face.

PROVING THE NEED;TO demonstrate the need for a new CRM, we must showcase its value and address the concerns surrounding change. Explain the limitations of the current system or processes and how the new CRM can address those pain points. Paint a picture of its benefits, such as increased efficiency, better customer insights, and improved collaboration. Share success stories and case studies of organizations that have seen a significant boost in productivity and revenue after implementing a CRM. Use real-world examples to demonstrate how the new system can streamline workflows, enhance customer experiences, and drive growth.

Improving Buy-In; Improving buy-in is about helping people understand and embrace the new CRM. Communication is key here! Be transparent about the reasons for the change and the potential impact it will have on individuals and the organization as a whole. Show empathy towards those reluctant to let go of their familiar ways. Offer comprehensive training and resources to ensure users feel comfortable and confident utilizing the new system. Assign dedicated champions or ambassadors who can share their positive experiences and guide others through the transition. Encourage questions and feedback, and be open to making adjustments based on user input. Remember, the more involved and empowered your team feels, the more likely they will embrace the change.

Potential Pitfalls; The road to adaptation is not always smooth. Some potential pitfalls may arise as you navigate implementing a new CRM. One common challenge is the resistance to change. People naturally tend to cling to what they know, fearing the unknown. To overcome this, create a culture of change readiness. Communicate the benefits and long-term vision of the new system, and involve users in the decision-making process whenever possible. Another potential pitfall is inadequate training. Ensure users receive proper training and support, addressing knowledge gaps or skill deficiencies. You can alleviate frustrations and help users adapt more smoothly by providing ongoing support and resources. Lastly, don't forget about the potential for data migration issues or integration challenges. Plan, conduct thorough testing, and involve key stakeholders to ensure a seamless transition and minimal disruption.

WHFN IMPLEMENTING A CRM SYSTEM, organizations face user resistance to change. This resistance often stems from the need to alter familiar routines and established habits. As a seasoned CRM professional, I have witnessed firsthand how this resistance can impede the success of CRM implementations. Addressing this issue head-on is vital, and implementing strategies to mitigate the potential negative impact.

Resistance to change can have far-reaching consequences,

including low user adoption, decreased productivity, and a reduced return on investment (ROI) for the CRM. To overcome this challenge, we must proactively implement improvements that promote user buy-in and facilitate a smooth transition to the new system.

One effective strategy involves convincing users of the benefits of the CRM system and ultimately mandating its use. This approach emphasizes the value proposition of the CRM, highlighting how it can streamline processes, improve efficiency, and drive better customer relationships. By clearly articulating the advantages the CRM can bring to their work, users are likelier to see the necessity and embrace the change.

Another essential improvement is providing comprehensive training and support to help users adapt to the new system. Change can be daunting, but with the proper guidance and resources, users can quickly become proficient in utilizing the CRM. By investing in robust training programs that cater to different learning styles and ongoing support channels, organizations can empower users to navigate the transition smoothly.

It is crucial to emphasize the benefits of adopting CRM throughout the training and support process. Users must understand how the CRM enhances daily operations, streamlines workflows, and contributes to individual and team success. By consistently reinforcing the advantages and demonstrating real-world use cases, organizations can foster a sense of excitement and motivation, encouraging users to embrace the change fully. Addressing resistance to change is vital for a successful CRM implementation. Organizations can increase user buy-in by convincingly articulating the benefits of the CRM system and establishing it as a mandate. Additionally, providing comprehensive training and ongoing support ensures users have the necessary skills and knowledge to adapt to the new system. Organizations can foster a positive transition and maximize user adoption by emphasizing the advantages of CRM and how it aligns with individual and organizational goals. Remember, patience, persistence, and the right strategies will pave the way for a successful CRM implementation and a high ROI.

. . .

REMEMBER, change is a journey that takes time, patience, and a touch of empathy. By proving the need, improving buy-in through effective communication and involvement, and navigating potential pitfalls proactively, you can set the foundation for a successful transition to your new CRM. Stay positive, be open to feedback, and remember that I'm here to guide you every step of the way.

9

The Importance Of Employee Training And Adoption In CRM Success:

Let's discuss something that can make or break a CRM implementation: proper training and employee buy-in.

IT'S ALL TOO easy to get caught up in the excitement of implementing a new CRM, but it's crucial not to overlook the importance of employee training and buy-in. Trust me; I've seen plenty of CRM implementations go sour despite having great features and capabilities because proper training and buy-in were not emphasized.

Proper training is essential to ensure employees are using the new system correctly. Even the most intuitive CRMs require learning and practice to use effectively. Without adequate training, employees may not know how to use essential features, leading to frustration, confusion, and mistakes. This can have a ripple effect throughout the organization, creating a hostile atmosphere and hindering the system's adoption. So, invest time and resources to properly train your employees, and trust me, it will pay off in the long run.

Of equal importance is employee buy-in. If employees don't see the value of the CRM, they aren't likely to use it, rendering all the investment and effort wasted. Employee buy-in can be achieved by clearly communicating the reasons behind the implementation and how the system will make their work easier and better, as well as highlighting the positive outcomes that the organization can realize. Encourage employee feedback and questions so they can voice their concerns and be included in the process. When employees feel heard and valued, it creates a more collaborative environment, and they are more likely to embrace the change.

As with any worthwhile endeavor, there are potential pitfalls to remember. First and foremost, invest enough time and resources in proper training so employees are equipped with the skills to use the system to its full potential. Failing to do so can create confusion, frustration, and low employee morale. Additionally, ensuring employees have access to the proper support channels (e.g., a helpdesk, FAQs, or a designated CRM expert) to receive guidance and support as they adapt to the new system. Finally, do not underestimate or undermine the importance of employee buy-in. It's essential to keep employees engaged and motivated throughout the implementation process. Remember, your employees are the ones who will be using the system day in and day out, so taking the time to win their trust and seeing things from their perspective is an investment that's worth making.

PROPER TRAINING and employee buy-in are crucial for several reasons:

Ensuring Effective Use Of The CRM; Employees who are well-trained in using the CRM are more likely to use it effectively, which leads to improved CRM, increased sales, and better decision-making. Proper training ensures that employees understand the full capabilities of the CRM and can make the most of its features and functionalities.

Increasing User Adoption; Employee buy-in is critical for successful CRM implementation, as it directly impacts user adoption. When employees understand the benefits of CRM and are confident in their ability to use it, they are more likely to embrace the new technology and incorporate it into their daily workflows.

Reducing Resistance To Change; Implementing a new CRM often involves significant changes to existing processes and workflows. Proper training and employee buy-in can help minimize resistance to these changes, as employees will be more likely to see the new system's value and be willing to adapt their ways of working accordingly.

Improving Data Quality; A well-trained workforce is more likely to input accurate and consistent data into the CRM, which is essential for maintaining data quality and ensuring the system delivers reliable insights and analytics. Employee buy-in also plays a role in data quality, as employees who are invested in the success of the CRM will be more likely to take the time to input data correctly and maintain data integrity.

Maximizing The Return On Investment (ROI); The ultimate goal of any CRM implementation is to maximize the ROI, which is achieved through improved customer relationships, increased sales, and more efficient processes. Proper training and employee buy-in are critical for achieving this goal, as they directly impact the effectiveness of the CRM and its ability to deliver the desired results.

WARNING: In CRM implementations, training is often an overlooked and neglected aspect of the project. When budgets are stretched, timelines are tight, and project fatigue sets in, training is frequently skimped upon. However, this is a perilous position because insufficient training on the new CRM and related processes can have severe consequences. It is critical to get the training right from the start.

It is essential to ensure that anyone who interacts with the system receives proper training, whether they input data, retrieve information, or make critical decisions based on CRM-generated insights. This includes leaders who rely on CRM-generated reports but may have limited knowledge of where the data originated. Neglecting their training jeopardizes the accuracy and integrity of the information they rely on for strategic decision-making.

Remember, a successful CRM implementation is not just about the technology; it's about empowering your people to make the most of it. Invest in comprehensive and customized training, and you will reap the rewards of improved efficiency, increased user adoption, and, ultimately, a transformative impact on your business's bottom line.

SUPPOSE a quality guide is not created and followed for the new CRM technology. In that case, several problems **will** arise, including:

Low user adoption; Without proper training and employee buy-in, user adoption of the CRM may be low, as employees may not understand the system's benefits or feel confident in their ability to use it. This can result in the CRM being underutilized, leading to a lower return on investment and reduced effectiveness in managing customer relationships.

Inefficient workflows; Employees not adequately trained in using the CRM may struggle to adapt their existing workflows and processes to the new technology. This can result in inefficiencies and reduced productivity, as employees spend more time navigating the system and less time on value-added activities.

Poor data quality; Employees may input inaccurate or inconsistent data into the CRM without proper training, negatively impacting data quality and the reliability of insights and analytics. This can lead to poor decision-making and reduced effectiveness in managing customer relationships.

Increased resistance to change; If employees are not

adequately trained and do not buy into the benefits of the CRM, they may be more resistant to the changes it brings. This can result in a longer implementation process, increased costs, and reduced ROI as the organization struggles to overcome internal resistance to the new technology.

Decreased customer satisfaction; If employees are not well-trained in using the CRM, they may struggle to provide the level of service and support that customers expect. This can decrease customer satisfaction, ultimately impacting the organization's bottom line.

Proper training and employee buy-in are critical for successfully implementing a CRM. By investing in comprehensive training and ensuring that employees understand the value of the new technology, organizations can maximize user adoption, improve data quality, and ultimately achieve a higher ROI. Failure to create and follow a quality guide can result in a range of problems, from low user adoption to decreased customer satisfaction, ultimately impacting the organization's ability to manage customer relationships and drive growth effectively.

Types of Training

RUNNING A BUSINESS IS HARD. Running a CRM to support a business can be even more complicated. We need to ensure we have the right people on board with the right skills and knowledge to run our CRM solution. That's why we have the training to cover all functional areas, including business process specialists, technical implementation specialists, and engagement managers. A training organization dedicated to supporting the CRM, its users, and customers should provide a comprehensive range of services, links, tools, and training materials to ensure successful implementation and ongoing proficiency. Here are some examples:

- **Training Programs:** Develop structured training programs tailored to the various user roles within an organization. These programs should cover all aspects of CRM usage, including system navigation, data entry, reporting and analysis, customer communication, and workflow management.
- **User Guides and Manuals:** Create user guides and manuals that serve as comprehensive references for CRM functionality. These resources should provide step-by-step instructions, screenshots, and explanations of key features, allowing users to quickly and independently resolve any issues they encounter.
- **Web-Based Training Modules:** Develop interactive and engaging web-based training modules that offer self-paced learning opportunities for users and customers. These modules should cover different CRM modules and features, allowing users to learn at their convenience.
- **Video Tutorials:** Produce a library of video tutorials demonstrating specific CRM tasks and processes. Users can easily access these tutorials, providing visual

guidance and enhancing their understanding of the system's capabilities.

- **Frequently Asked Questions (FAQs):** Create an online repository of FAQs that address common queries and challenges related to CRM usage. This resource should be regularly updated to incorporate new questions and user feedback.
- **Knowledge Base:** Establish a comprehensive knowledge base that documents best practices, troubleshooting tips, and advanced configuration options. This knowledge base should be easily searchable and regularly updated to ensure quality support and up-to-date information.
- **Discussion Forums and Community Platforms:** Set up online forums or community platforms where users and customers can interact, share experiences, and seek assistance. Moderators should actively participate in these forums to answer questions and foster a collaborative environment.
- **Online Help Desk:** Implement an online help desk system to collect and manage support requests from users and customers. This tool should enable efficient ticket tracking, escalation, and resolution with clear communication channels.
- **Proactive Outreach:** Conduct regular webinars, workshops, and training sessions to proactively engage users and provide updates on new CRM features, best practices, and industry trends. These initiatives should encourage ongoing learning and foster a sense of community among CRM users.
- **Integration Guides:** Produce integration guides that detail how the CRM can be seamlessly integrated with other applications and platforms commonly used by customers. These guides should provide step-by-step instructions, known issues, and troubleshooting tips.

- **Release Notes and Product Updates:** Regularly communicate product updates, enhancements, and bug fixes to users and customers via release notes. These documents should provide a clear overview of changes, their impact, and any additional steps users require.
- **Customer Success Managers:** Assign dedicated customer success managers who can provide critical customers personalized guidance, training, and support. These managers should have in-depth knowledge of CRM and work closely with customers to ensure their success.

A training organization can empower CRM users and customers by offering comprehensive services, links, tools, and training materials, enabling them to maximize the system's benefits and meet their specific business needs.

10

Internal User Adoption

The most common explanations for user resistance to CRM implementation and usage usually fall into two areas:

The most common reasons for user resistance to CRM implementation and usage usually fall into a few places:

- Lack of support by a potential user's leadership
- Perceived inconvenience usually manifests itself by the perceived thought that using the CRM will cause more and/or redundant manual work.
- Concern for intellectual property theft
- Fear of excessive productivity monitoring
- Reluctance to change and adapt

Lack of Support by a Potential User's Leadership:

One of the most daunting challenges in CRM implementation is addressing employee resistance to change. This resistance can be particularly prominent among leaders who are comfortable with their current methods and processes and are unwilling to consider a new way. This issue becomes even more complex when leaders have developed an existing system, either companywide or for their area of responsibility, and have become emotionally attached to this pre-existing system and process. In such cases, leaders may feel professionally threatened by introducing a new CRM system.

I have encountered these challenges repeatedly. Handling the situation delicately and with finesse is crucial to address resistance among leaders. This must be handled with kid gloves.

One practical approach is to **incorporate the resistant leader into the new CRM process** and "flip" their resistance to champion the new effort. By involving them in the planning and decision-making stages, you can help them understand the benefits of the CRM system and how it can improve efficiency, productivity, and decision-making capabilities. This helps overcome their attachment to the existing processes but also helps alleviate their concerns about professional threats. As they become more engaged and see the value of the CRM, they are more likely to advocate for the new system and actively drive its implementation.

Another strategy to address resistance among leaders is to **prove the need for the CRM system**. Remember that if you choose this path, the leader will likely assume you are a "used car salesman," you will start this journey in a ditch alongside the road and have to pull yourself out very slowly. If you don't heed this warning, be prepared to be hit by a big truck and not back into a deeper ditch - I warned you! Many leaders may resist change because they do not fully understand or appreciate the benefits that CRM can bring to the organization. It is essential to present them with compelling evidence, such as case studies, testimonials, or data-driven analyses, demonstrating how the CRM can improve efficiency, productivity, and decision-making capabilities. By showcasing real-life examples of the positive impact that CRM has had on similar organizations, you can help leaders understand the value of adopting a new system and overcome their resistance.

Improving buy-in among leaders is another crucial step in addressing resistance to CRM implementation. Involving the leaders in the implementation process gives them a sense of ownership and empowerment over the new system. This can go a long way in transforming their resistance into enthusiasm and support for the CRM project. **Soliciting their input, ideas, and feedback** demonstrates that their perspectives are valued and respected. By incorporating their ideas and suggestions into the CRM implementation plan, leaders are more likely to feel invested in the project and become champions of change within the organization.

However, it is essential to acknowledge potential pitfalls if leaders' resistance is not addressed effectively. If leaders remain attached to the existing system and processes, it will impede employee adoption of the CRM system. This can result in low user adoption rates, decreased productivity, and a reduced return on investment (ROI) for the CRM. If leaders are not on board and actively supporting the implementation, it will create a ripple effect throughout the organization, influencing how employees perceive and engage with the new system. Hence, it is crucial to address the concerns and resistance of leaders to ensure the successful adoption and utilization of the CRM system.

Everyone must approach the situation with caution and finesse when they resist implementing a new CRM system. By incorporating resistant leaders into the CRM process, demonstrating the value and benefits of the system, and involving leaders in the implementation process, organizations can overcome resistance and promote the adoption and utilization of the CRM. However, it is essential to be aware of the potential pitfalls if resistance is not effectively addressed. By proactively addressing opposition and gaining the support of leaders, organizations can maximize the benefits of the CRM system and drive positive change throughout the organization.

WARNING: Addressing the challenges of lack of support by potential users' leadership and perceived Inconvenience is essential for ensuring the successful adoption of a CRM. It is ubiquitous and frequently hidden as the true culprit. By proving the benefits of the CRM, involving resistant leaders in the implementation process, and demonstrating the time-saving capabilities of the system, organizations can overcome these obstacles and achieve greater user adoption, increased efficiency, and improved decision-making. Ignoring these challenges, on the other hand, can result in decreased user adoption, reduced ROI, and impaired growth. Approach these challenges with professionalism and insight, and enjoy the benefits of a well-adopted and optimized CRM for years to come.

Perceived Inconvenience:

User resistance is among the most significant limitations, particularly from on-the-go salespeople and professional services personnel. These users tend to view any change, especially such a significant shift as implementing a CRM system, as a major inconvenience and, in some cases, a pain in the ass. Resistance from these core users is not uncommon and should be taken seriously as it can cause significant waves in the implementation project. It is crucial to anticipate and debunk their reservations through suitable and appropriate means.

To address user resistance in general and salespeople's and professional services personnel's reservations, it is crucial to identify the root causes. One of the primary reasons for resistance is the perceived inconvenience of CRM implementation. Users often cite that adoption requires learning a new system, which they don't have time to do or deem unnecessary. Such reservations and resistance should not be overlooked as they can directly impede adoption, ultimately impacting the success and ROI of the CRM initiative.

The issue of user resistance must be debunked through various strategies. One is to showcase and communicate the CRM's benefits to all users, particularly its capability to improve their work lives. For salespeople and professional services people, the CRM can facilitate their day-to-day tasks, improve the accuracy of their customer interactions, and deliver valuable insights into managing customer outcomes. By highlighting the personal benefits to the users, organizations can create a more positive perception of the CRM's potential.

Another strategy is to emphasize the CRM's ease of use and convenience. Users should be assured that learning and getting familiar with the system is easy and will not take too long. Interface design that is intuitive for technical and non-technical users, quality training sessions, and readily available support systems are all vital to improving the learning experience. Time constraints of users should also be evaluated, whereby training can be re-designed and broken down into smaller chunks that can be fitted more easily into their busy schedules.

Further, organizations can implement various methods to make using a CRM system part of the users' lifestyle. A CRM-optimized interactive interface can be designed to sync with users' preferred mobile devices giving them the freedom and flexibility to work on the go. Making use of the CRM work with existing tools is also ideal. Integration with the user's Outlook, calendar, and email system make their work more efficient without making them feel like it is a burden.

It is vital to create personal connections with these users as well. It helps to have someone on the team dedicated to personalized follow-up with particular attention to those with reservations. They can address their concerns, including answering questions, and arrange real-time demonstrations or case studies demonstrating success with other clients. Providing additional support and specific training to users with a demonstrated need will create positive momentum for the initial adopters and foster a culture of acceptance for the wider group.

However, it is essential to note that not addressing reservations and user resistance can lead to detractors who, in turn, influence the opinion of other users in the setting. This can slow adoption and lead to suboptimal use of the CRM, hence lower ROI. It is thus crucial to anticipate and address reservations proactively and purposefully to manage and develop these detractors into advocates for the CRM system.

Resistance from on-the-go users can impede adoption, causing user waves that can affect the CRM project's overall success. Organizations should have strategies tailored to anticipate and debunk user reservations through a multi-pronged approach to address the perceived inconvenience of adopting a new CRM system. By delivering concrete assurances of personal benefits, ease of use, improvement of lifestyle, personalized follow-up, and the establishment of advocates, organizations can speed up adoption, enhance engagement, and improve CRM ROI.

When it comes to adopting a CRM system, one significant

hurdle organizations often face is the perception of inconvenience among users, especially those constantly on the go, such as salespeople or consulting professionals. Adding more administrative work to their busy schedules can be a death sentence for any CRM effort. Therefore, It is crucial to create an understanding and perception that the CRM is a time-saving tool that will ultimately reduce their administrative burden. This perception of time-saving benefits must be repeatedly proven and reinforced to gain the users' and their leadership's trust and buy-in.

To address the perception of inconvenience, it is essential to provide concrete evidence of the time-saving potential of the CRM system. One way to achieve this is by conducting pilot tests or simulations that showcase how the CRM can streamline administrative tasks, automate manual processes, and improve overall efficiency. By sharing the results of these tests with users and their leadership, you can build confidence in the CRM as a valuable time-saving tool. When users and their leaders see tangible evidence of the time saved through the CRM, they are more likely to embrace it and actively support its implementation.

Improving the user experience is another critical factor in addressing the perception of inconvenience. Designing the CRM with user convenience is essential, incorporating a user-friendly interface, simplifying navigation, and customizing the system to fit users' specific needs and preferences. By tailoring the CRM to align with the unique requirements of the users, you can minimize the perceived burden of adoption and make it easier for them to integrate the system into their daily routines.

In addition to design considerations, providing comprehensive training and ongoing support is crucial to help users become proficient with the CRM and maximize its time-saving capabilities. Training sessions should focus on the technical aspects of using CRM and highlight how the system can streamline its workflows and improve efficiency. Ongoing support should be readily available to address any questions or challenges users may encounter during their CRM journey. By investing in proper training and support, you can mitigate the barriers that users may face and enhance their overall experience with the CRM.

It is essential to recognize the potential pitfalls if the perceived inconvenience is not adequately addressed. Suppose users continue to view the CRM as a time-consuming burden. In that case, it can lead to low user adoption rates, decreased productivity, and a reduced return on investment (ROI) for the CRM. Without addressing these concerns, organizations risk undermining the success of their CRM initiative and missing out on the many benefits that a well-implemented CRM system can provide.

Addressing the perception of inconvenience among users is crucial for successfully adopting and implementing a CRM system. Organizations can overcome the hurdles associated with user resistance by demonstrating the time-saving benefits through pilot tests or simulations, designing a user-friendly interface, and providing comprehensive training and support. Failing to address the perception of inconvenience can result in low user adoption, decreased productivity, and diminished ROI for the CRM. By proactively addressing these challenges, organizations can maximize their CRM system's potential and drive significant efficiency and productivity improvements.

Intellectual Property Theft:

Some users may resent it because they fear the CRM will steal their intellectual property. They worry that their hard-earned insights, sales tricks, and valuable contacts may be snatched away only to benefit the organization. This concern is rooted in the misconception that CRM solely focuses on data collection without considering the security and privacy of individual users' intellectual property.

Remember, clear communication, education, and addressing concerns are crucial to addressing the matter of intellectual property theft and ensuring the successful adoption of a CRM. Emphasize the benefits, involve users in the process, and proactively address potential pitfalls to ensure a smooth and positive experience.

Addressing the concern of stealing a person's intellectual content by the CRM and exploring how we can prove the need, improve buy-in, and navigate potential pitfalls in this situation.

PROVING THE NEED; To address the concern of intellectual property theft, it's essential to demonstrate the purpose and benefits of the CRM. Build a case for how the system can enhance users' work and protect their intellectual property. Emphasize that the primary goal of the CRM is to organize and centralize valuable information, making it more accessible and secure. Highlight the data encryption and security measures to safeguard sensitive information. Back up your claims with success stories or case studies, showcasing how CRMs have been instrumental in protecting valuable intellectual property and driving business growth.

Improving Buy-In; Improving buy-in from users requires effective communication and education. Take the time to explain the processes and technology behind the CRM. Provide training and resources that help users understand how the system works and how it benefits them. Address their concerns directly and debunk any misconceptions. By fostering an open dialogue and involving users in decision-making, you can build trust and encourage their active participation. Show how the CRM can streamline workflows, improve collaboration, and ultimately enhance productivity and success. Additionally, consider appointing ambassadors or champions who have embraced the CRM and can share their positive experiences with their peers.

Potential Pitfalls; When it comes to potential pitfalls, it's essential to address them proactively and have a game plan in place. One potential pitfall is the accidental exposure of sensitive information. Establish access controls and permissions to mitigate this risk and ensure only authorized individuals can view and manipulate specific data. Regularly review and update security protocols to stay ahead of potential threats. Another pitfall to watch out for is resistance to change. It's natural for users to resist new technology, especially if it threatens their familiar way of doing things. To navigate this, provide comprehensive training, hands-on support, and ongoing assistance to ensure a smooth transition. Encourage feedback and adapt the CRM based on user input to address pain points and make it more user-friendly. Lastly, be mindful of data ownership rights and communicate how the intellectual property of individuals will be protected within the CRM.

11

Overcoming Resistance To CRM Adoption:

Examine employee resistance to CRMs and strategies for addressing these concerns to ensure system success. Employee resistance to new CRM systems can be a thorny issue when implementing and adopting this technology. It's important to dive deep into the reasons behind this resistance and develop effective strategies to address these concerns. Ensuring the success of the CRM depends on the buy-in and engagement of your employees.

So, why do employees sometimes resist new CRMs? - There can be a variety of reasons. Some employees may fear change and be comfortable with the existing systems and processes they are familiar with. Others may worry about the learning curve and the potential disruption to their day-to-day work. Some may not understand the benefits of the new CRM and how it can make their job easier and more efficient.

To successfully address these concerns, it's crucial to communicate the value and benefits of the new CRM to your employees. Clearly explain how it will enhance their work and make their lives easier. Share success stories or case studies from other teams or departments who have already adopted the CRM and are reaping the rewards. Additionally, involve employees in decision-making as much as possible, seeking their input and feedback. This will make them feel valued and more invested in the transition.

Another effective strategy is to provide comprehensive training and support. Offer hands-on workshops, online tutorials, or even one-on-one coaching sessions to help employees familiarize themselves with the new CRM. Ensuring they have the necessary skills and knowledge to use the system effectively alleviates their fears and empowers them to embrace the change.

Addressing any resistance head-on is also essential. Encourage open and honest communication with your employees. Listen to their concerns and address them proactively. Provide a platform for everyone to ask questions, share their challenges, and provide feedback. Creating a supportive environment can help employees overcome resistance and develop a sense of ownership and enthusiasm for the CRM. Celebrate small milestones and achievements along the way. Recognize and reward employees who demonstrate proficiency and engagement with the CRM. Acknowledging and appreciating their efforts foster a positive atmosphere and motivate others to get on board.

Employee resistance to new CRMs is not uncommon. Still, understanding the reasons behind this resistance and implementing effective strategies to address concerns can pave the way for the successful implementation and adoption of the technology. Ultimately, by bringing your employees along on this journey, you'll be maximizing the potential of your CRM and setting your business up for long-term success.

12

Some Reasons Behind Employee Resistance To CRMs

When it comes to implementing a new CRM system, it is not uncommon for employees to exhibit resistance. Understanding the reasons behind this resistance is crucial for project teams and administrators to address the concerns of their workforce effectively. Employee resistance can stem from various factors, including asshole leaders, scared leaders, fear of change, lack of understanding of the benefits, concerns about job security, inadequate training, and a perceived increase in workload.

Assholes, unfortunately, can hinder the successful implementation of a CRM. These individuals, usually leaders, may have initially pledged to champion the system but failed to fulfill their commitment to the project team. Their lack of support and active resistance to the new CRM sends a negative message throughout the organization and undermines the project's credibility. Asshole leaders may have personal agendas or refuse to accept change. This can create a hostile environment and make it challenging for employees to embrace the new system.

Similarly, scared leaders can also impede the implementation process. These leaders are typically resistant to change because they may not fully understand the benefits or implications of the CRM system. They may fear that the new technology will diminish their importance or authority within the organization. This fear can result in an unwillingness to champion the system, leading to a lack of endorsement and support from higher levels of management. Overcoming the apprehension of such leaders is crucial for ensuring a smooth transition and gaining widespread acceptance among employees.

Fear of change is a typical response among individuals when introducing a new CRM system. Implementing a CRM often requires significant changes to existing processes and workflows. Employees may be concerned about how these changes impact their daily tasks and routines. They may fear a loss of efficiency or worry that the new system will disrupt established working methods. This fear of change can manifest as resistance to the new CRM, as employees try to hold onto the familiar and resist the unknown. Addressing this fear requires clear and transparent communication about the benefits and advantages of the CRM system. It is essential to emphasize how the new system will streamline processes, improve efficiencies, and make their work easier.

Lack of understanding of the benefits is another significant factor contributing to employee resistance. Employees who do not fully grasp how CRM can improve their work and make their lives easier may hesitate to embrace the new technology. Project teams and administrators must invest time and effort in clearly communicating the advantages of the CRM system. This can include providing demonstrations, training sessions, and sharing success stories from other departments or organizations successfully implementing CRM. Ensuring employees see the value in adopting the CRM is crucial for overcoming resistance and fostering a positive attitude towards the new system.

Concerns about **job security** can also contribute to employee resistance to CRM implementation. Some employees may fear the CRM system will automate tasks or render specific roles obsolete, potentially leading to job losses. This fear can be compounded by rumors or miscommunication about the impact of CRM on job roles within the organization. Addressing these concerns requires open and honest communication about the goals and intentions of the CRM implementation. Assuring employees that the purpose of CRM is to enhance their work and improve customer relationships rather than replace them can help alleviate their fears and resistance.

The level of training employees provide plays a crucial role in their acceptance and adoption of the new CRM system. **Inadequate training** can leave employees feeling overwhelmed or ill-equipped to navigate the system effectively. This lack of confidence can lead to frustration with the CRM and a preference for sticking to their familiar, existing processes and tools. Proactive training programs tailored to the specific needs of different user groups are essential for building competence and confidence in using CRM. Project teams and administrators can mitigate resistance and encourage adoption by ensuring employees feel adequately prepared and supported using the new system.

Lastly, employees may resist the implementation of a CRM due to their **perception that it will increase their workload**. This perception can arise from the belief that the CRM will require additional data input or that employees must learn new processes on top of their workload. This perceived increase in workload can make employees resistant to the new system, as they may view it as burdening rather than facilitating their work. Addressing this perception requires clear communication about the long-term benefits of CRM, such as reduced manual data entry, streamlined processes, and improved efficiency. Additionally, providing the necessary support and resources, such as dedicated training and ongoing technical assistance, can help alleviate concerns about an increased workload.

Understanding the reasons behind employee resistance to CRM implementation is crucial for project teams and administrators. Asshole leaders, scared leaders, fear of change, lack of understanding of the benefits, concerns about job security, inadequate training, and a perceived increased workload can all contribute to employee resistance. Addressing these concerns requires effective communication, comprehensive training programs, and proactive support to ensure a smooth transition and foster acceptance of the new CRM system. By acknowledging and overcoming these challenges, project teams and administrators can maximize the benefits of CRM for the organization as a whole.

13

Strategies For Addressing Employee Resistance To CRM

Implementing a new CRM system can be difficult, especially when faced with employee resistance. Addressing and overcoming resistance is crucial for the success of the implementation, ensuring that the organization can fully leverage the benefits of the CRM. Organizations can adopt several strategies to combat resistance and promote acceptance and adoption. Let's dive into each strategy in more detail:

Identify Difficult Leaders; one of the first steps in addressing employee resistance is identifying and addressing difficult leaders within the organization. Difficult leaders can hinder the successful implementation of a CRM system, as their lack of support and active resistance sends a negative message throughout the organization and undermines the project's credibility. It is important not to isolate these leaders but to retrain and comfort them. By providing them with the necessary support and education, they may be more likely to become advocates for the CRM system and help drive its success.

Communication Of CRM Benefits; one of the most effective strategies for addressing employee resistance is clear and compelling communication of the benefits of the CRM system. It is crucial to articulate to employees how CRM will improve their daily work and contribute to the organization's overall success. Take the time to explain how the CRM will streamline processes, enhance customer relationships, and enable employees to focus on higher-value tasks. By helping employees understand the value and impact of CRM on their work, they are more likely to embrace the technology and actively support its implementation.

Employee Involvement In The Implementation Process; involving employees in the planning and implementation of the CRM system is another practical approach to combat resistance. Seek their input and feedback on the CRM and its features, and allow them to shape the system according to their needs and requirements. This involvement creates a sense of ownership and buy-in among employees, making them more likely to support the new system. Additionally, involving employees in the decision-making process fosters a sense of empowerment and can help to alleviate fears and concerns about the CRM.

Comprehensive Training Program; one of the leading causes of resistance is a lack of understanding or confidence in using the CRM system. To overcome this, providing employees with comprehensive training on using CRM effectively is crucial. The training program should be tailored to employees' needs and roles, considering their existing knowledge and skillset. Hands-on practice and ongoing support should be provided to help employees become comfortable with the new system. By investing in comprehensive training, employees will feel better equipped to embrace CRM and incorporate it seamlessly into their daily work.

Addressing Job Security Concerns; a common problem among employees is the fear that the CRM will render their roles obsolete and lead to job losses. Addressing these concerns is critical for addressing resistance. Reassure employees that the CRM will enhance their positions and make their work more efficient rather than replace them. Emphasize that the CRM will enable employees to focus on higher-value tasks and improve their skills and productivity. By communicating the long-term benefits and positive impact of CRM on job security, employees are more likely to embrace the system and feel more secure in their roles.

The Gradual Introduction Of The CRM; instead of implementing the CRM system all at once, consider introducing it in stages. This approach can reduce resistance by giving employees time to adjust to the new system and processes. Start by implementing one aspect of the CRM system and allow employees to become familiar with it before moving on to the next. This gradual approach minimizes disruption and will enable employees to adapt to the changes gradually. By easing employees into the new system, resistance can be reduced, and employees can become more comfortable with the CRM over time.

14

Remote Access

Warning: I can attest that mobile access concerns are common among users. However, this fear is often unfounded, as modern CRM systems offer far greater mobile capabilities than typically achieved without a CRM. I urge you to proactively address this misconception by communicating the extensive mobile features of your CRM system through compelling training programs, documentation, and other targeted communication channels. By emphasizing the benefits of mobile access, including enhanced productivity, collaboration, and responsiveness, you can inspire your users to embrace this tool as a valuable asset in their CRM journey. Remember to approach this concern with a cautionary tone and stress the importance of due diligence when dispelling misconceptions and empowering users to leverage the full potential of their CRM system while on the go.

WHEN USERS FIND themselves without access to important CRM data while away from the office, it can pose significant challenges to their ability to work effectively. This lack of access or the perception thereof can lead to frustration, decreased productivity, and missed opportunities. However, addressing this issue pragmatically and dispelling any misconceptions is essential.

While it may seem like a significant limitation, **the truth is that modern CRM systems have evolved to offer robust mobile and remote access capabilities.** These capabilities enable users to access important information wherever they may be - whether on the road, at a client's office, or even from the comfort of their homes. It's essential to emphasize that this is not just a gimmick or a luxury feature but a critical element that can enhance productivity and drive business success.

I strongly encourage you to harness all of the mobile capabilities of your CRM to their fullest extent. However, we must maintain a cautious approach and recognize the potential pitfalls when utilizing remote access. Allow me to guide you through this process with the benefit of my experience.

First and foremost, let's emphasize the importance of security. While remote access provides convenience and flexibility, it also introduces additional security risks. Ensure your CRM system has robust security measures, such as strict user authentication protocols, encrypted data transmission, and proper access controls. By prioritizing security, you can safeguard sensitive customer data and protect your organization from potential breaches.

Next, let's consider the ease of data entry. Remote access should not compromise the accuracy and completeness of data entered into the CRM system. Educating your team on the proper data entry practices is essential, and emphasizing their responsibility when inputting information remotely is critical. Conduct regular training sessions and provide resources highlighting the significance of maintaining data integrity.

One significant advantage of remote access is its immediate access to information. This allows your team to retrieve crucial data in real time, enabling them to make informed decisions regardless of their physical location. However, be mindful of potential connectivity issues or data synchronization challenges that may arise. Perform thorough testing to ensure seamless access to up-to-date information on any device and from any location, thus minimizing the risk of making decisions based on outdated or incomplete data.

Lastly, due diligence is critical when implementing and utilizing remote access features. Conduct a comprehensive review of the compatibility of your CRM system with various remote devices and operating systems. Test the functionality and performance of the system across different scenarios to proactively identify and address potential issues. Additionally, watch for updates and advancements in remote access technology to ensure your CRM system remains aligned with industry standards and requirements.

THE REMOTE ACCESS capabilities of your CRM system offer great potential for increased productivity, efficiency, and access to real-time information. It's vital to approach this feature cautiously and prioritize security, data entry accuracy, and due diligence. Taking these precautions allows you to leverage the full benefits of remote access while minimizing potential risks. As a seasoned professional, I share this wisdom with you to help you achieve the best outcomes for your CRM implementation.

15

Notification/Alerting System

It would be best if you placed high importance on having an effective alerting system within your CRM. Without this valuable tool, users may not feel comfortable storing and sharing information, leading to decreased productivity, missed deadlines, and even loss of business opportunities. Taking a pragmatic approach to this concern is crucial, and offering actionable solutions ensures users have the tools to stay informed and on top of their tasks.

Developing a robust alert system within the CRM should be a priority for all involved parties. An effective alert system should give users timely notifications for activity related to clients, use of intellectual property (IP), important departmental events, deadlines, and tasks. These alerts should be highly configurable to meet specific needs, enabling users to adjust the warning system to fit their concerns, schedules, and priorities. By allowing users to customize the warning system to their specific needs, you can empower them to be more productive and engaged in their work.

Understanding the potential consequences of not having an effective alerting system is essential. Losing even one business opportunity or losing the confidence of a key star-performer could significantly impact an organization's success. Therefore, it is imperative to create a comprehensive alert system within the CRM and provide training and demonstrations to users to ensure they are comfortable using the new system.

To help users feel more comfortable regarding the new system, live demonstrations and assurances regarding the reliability of these features should be provided. Bringing users onto the new system in person gives them a sense of comfort and enables them to experience the new system's benefits firsthand. Additionally, proper training and ongoing support are essential to ensure users understand how to use the alerting system effectively.

A robust alerting system is essential for any successful CRM implementation. By creating a highly configurable warning tool and providing training and demonstrations to users, you can help them avoid missing critical information, increase productivity, and prevent the loss of business opportunities. As a mentor-like guide, I stress the importance of due diligence when addressing this concern and providing actionable solutions to ensure users have all the necessary tools for a successful CRM experience. Remember, managing any potential pitfall requires a pragmatic approach, and addressing this concern is no exception.

16

Finding The Correct Information

I understand the frustration and inefficiency that can arise when users consult various sources to find necessary information. This issue can lead to user dissatisfaction, low adoption rates, and decreased productivity - ultimately impacting the effectiveness of the entire system. Rather than dismissing this as an excuse, it is essential to tackle this issue head-on and guide users on more efficient information retrieval practices.

In our journey towards successful CRM adoption, it's essential to recognize that users may not immediately grasp the immense potential and possibilities that come with the newly available knowledge and information available to them. Showing empathy and understanding towards users during this transition is paramount. We must acknowledge their willingness to try and their genuine desire to make the system work. Our role as professionals is to guide them through this process and help them overcome any hurdles or hang-ups they may encounter.

Users often get hung up on seemingly minor details that hinder their system adoption. It's crucial to differentiate between complaints about actual problems that require remediation and concerns that stem from a lack of familiarity with the system's full capabilities. Recognizing this distinction allows us to approach the situation with patience, empathy, and experience gained from countless CRM implementations.

Education becomes crucial to address the perspective that users cannot be expected to grasp the wealth of knowledge and possibilities immediately. We must proactively educate users on the value of accessing more insightful, direct, and up-to-date information. By emphasizing the importance of real-time data and highlighting the benefits of not having to request reports or wait for compiled dashboards regularly, we can help users see the true power of the CRM system.

Empowering users with tools and encouraging them to explore the system more proactively can make a difference. Providing training sessions, resources, and ongoing support enables users to adapt and retrieve the necessary information efficiently and quickly. Sharing real-life examples and success stories from other departments or companies can also inspire users and demonstrate the potential positive impact of the system.

However, it's essential to remember that fostering adoption and overcoming user resistance is ongoing. Even after initial training, users may struggle to fully embrace the system's potential. Continuously reinforcing the benefits, sharing success stories, and providing avenues for ongoing education and skill development can help users gradually unlock the possibilities of this newfound wealth of knowledge.

Remember, our role is to implement the CRM system and guide users through the mindset shift required for its successful adoption. By approaching this with empathy, patience, and mentorship, we can make a difference in helping users understand and leverage the vast potential of the CRM system.

Let's navigate this challenge with the benefit of our experience, always stressing the importance of due diligence and highlighting potential pitfalls in a cautionary tone.

Another crucial step is training users on how to work for themselves and navigate their way around the CRM system. This includes ensuring that users know where the necessary information resides within the CRM and how to use the system's search functionality effectively. Improving search functionality is also critical to making information retrieval more efficient. With intuitive search capabilities, users can easily find the relevant information they need, reducing frustration and saving time.

Ensuring that the support organization is adequately staffed, publishes SLAs for support, and is empathetic and responsive is also essential. This support structure is necessary to assist users when they encounter difficulties in the system - providing guidance, troubleshooting, and support to help users achieve their goals. By having trustworthy and responsive support available, users can gain confidence in the system and be more willing to explore. Inefficient information retrieval can impact user satisfaction and productivity. It is essential to educate users on the value of accessing more insightful information, empower them with the necessary tools, and provide adequate training on using the system effectively. By improving search functionality and ensuring support structures are in place, we can mitigate the negative consequences of inefficient information retrieval and create a more efficient and effective CRM system.

I have seen firsthand the benefits of adopting these practices - and I encourage others in the field to do the same. Remember, you're not just sharing knowledge; you're guiding your reader through potential challenges using the benefit of your experience.

17

Insufficient Reports

I've encountered the complaint of insufficient reports within CRM systems many times throughout my career. This caught me off guard, too, as CRMs are not usually considered to be lite in the reporting of the information department. To this point precisely, it is crucial to realize that **there is a big difference between the ability to report more information and the actuality of doing so.** However, addressing this concern requires a pragmatic mindset, which can indicate a breakdown in the business analyst's work or communication between the CRM team and the users. I advise this situation requires a thorough analysis of the root causes to ensure a successful CRM reporting experience.

Users who perceive the lack of readily available reporting tools within a CRM system may find it challenging to generate meaningful insights and make data-driven decisions. This perception can lead to frustration, low adoption rates, and impaired decision-making. Therefore, it's essential to understand the context of the situation and focus on providing customizable, ready-to-use reports and dashboards that empower users to generate insights and make informed decisions quickly.

If this is a widespread concern among users, it indicates an issue with the development team. It's crucial to address this with the team and ensure they enhance the CRM's reporting capabilities while listening adequately to the customers' needs. Continual education is also vital in reporting capabilities. Never stop educating users about the existing reports, and always consider feedback and suggestions for new reports. By involving users in the process and understanding their requirements, you can ensure that the reports you develop are relevant and valuable for multiple departments or teams.

Creating a policy with precise criteria for report creation is essential to maintain focus and prevent the creation of unnecessary or redundant reports. Reports should only be made if they are relevant to a certain number of departments or teams, and one-off reports should be avoided unless others could benefit from a similar report. Another approach is internally allocating funds or charging for accurate one-off reports or ASAP requests, incentivizing users to consider the reporting needs carefully.

Consistency is vital when implementing policies or allocating funds for reports. By doing so, users understand that reporting resources should be used judiciously, reinforcing the importance of due diligence. Addressing the concern of insufficient reports in the CRM system requires a thoughtful analysis of the situation and strategic interventions. Enhancing the CRM's reporting capabilities, actively educating users, adopting a proactive approach to report development, and implementing clear policies can alleviate user frustration, improve adoption rates, and enable better decision-making.

I stress the importance of being proactive and ensuring users have the tools and resources they need for a successful CRM reporting experience. Remember, with careful planning and execution, you can turn the frustrating issue of strict reporting capability into a valuable and productive asset.

WARNING: If reports are genuinely unavailable within your CRM system, it is critical to take immediate action. I have encountered

this issue before and learned the importance of conducting a thorough review to bridge this gap swiftly and effectively.

To begin, I recommend initiating a side-project dedicated to reviewing the reports that are currently available, understanding their purpose, identifying their intended recipients, and analyzing the disparity between what people or departments initially requested, what they had access to before implementing the new CRM system, and what they currently expect.

By embarking on this side project, you can gather valuable insights into the underlying causes of the "missing" reports and develop a comprehensive plan to fill this void promptly. Dedicate the necessary resources and assemble a cross-functional team consisting of business analysts, CRM specialists, and key stakeholders from the affected departments.

It's crucial to actively engage with users during the review process to gather their input and requirements. By involving them in the analysis, you can better understand their reporting needs and expectations. This will inform the development of a roadmap to address the missing reports and ensure they align with the broader goals of your CRM system.

As you identify gaps and discrepancies, consider the reasons behind them. It could be limitations within the CRM system or a lack of understanding about utilizing its reporting capabilities. Once these issues are identified, prioritize them for resolution and develop an actionable plan to address each one.

Remember, time is of the essence. Promptly bridging the gap in missing reports will alleviate frustration, improve user adoption, and enable evidence-based decision-making. With a dedicated "side" project, a multidisciplinary team, and a comprehensive roadmap, you can swiftly tackle this issue and unlock the full potential of your CRM system.

18

Case Study: Duplicative Reporting To The Boss

Scene:

Picture a company striving to streamline its sales process by implementing a modern CRM system. This system is intended to be used by ten similar business industry groups , each with existing best practices and lightweight tools like Excel to report information. The organizational structure is built like a pyramid, with each industry group having 30 VPs reporting to their group leader.

Additionally, the leader of each BIG has unique reporting requirements, and in turn, each BIG may have to submit yet another report to the company's overall leadership. This labyrinth of reporting may seem amusing, but it sets the stage for a critical incident.

Incident:

The incident revolves around adherence to the new CRM. If VPs, group leaders, or company leaders are not entirely on board with the new system, they may continue using their old methods and tools.

Picture a scenario where a lack of adherence to a new CRM poses a significant challenge. VPs, group leaders, or company leaders may continue using their old tools and methods, hindering the CRM's effectiveness.

This lack of compliance is a significant problem that can jeopardize the entire CRM implementation. It's crucial to identify each process, address non-compliance, and secure compliance across the reporting pyramid, from top to bottom.

In my experience, securing buy-in from key leaders is critical to ensuring success in any CRM implementation. Without the support of those in high positions, it's challenging to get teams to adopt a new system fully.

It's essential to take a strategic approach, set clear expectations, and communicate why the new system is necessary to drive organizational growth. Training, support, and incentives for using the new system can help overcome resistance and ensure widespread adoption.

My mentor-like advice to you is to approach your implementation with due diligence, focusing on building a comprehensive strategy that fosters collaboration and drives the adoption of the new system. Empower your teams by providing them with the means to adapt to the new system and ensure compliance across all levels of the reporting structure.

Remember, the key to success is choosing the right CRM and ensuring it is fully adopted across the organization. Taking the time to secure buy-in from key leaders and nurturing a culture of adoption and compliance is critical.

Result:

Non-compliance and the deliberate use of the old system, even on a small scale, will negatively impact the success of the entire CRM in two primary ways:

Resentment and Duplication; Individuals who need to use old and new systems will resent the duplicative work and inefficiency. This frustration can decrease morale and productivity as employees struggle to navigate multiple tools and processes.

Diminished System Reliability; The reliability of the entire CRM, or the perception of its reliability, will suffer as duplicative work introduces data lag. Inconsistencies and delays in data input can erode trust in the system, undermining its effectiveness and creating unnecessary stress for both the people and the technologies involved.

Lessons Learned:

The incident unveils several critical lessons to consider while implementing a modern CRM:

Secure Buy-In From All Levels; It is vital to gain the support and commitment of key stakeholders, including VPs, group leaders, and company leaders. Ensuring alignment and buy-in will minimize non-compliance risk and increase the system's overall success.

Identify And Communicate Clear Transition Dates; Communication is vital in managing change. By establishing a precise date for the complete transition from the old to the new system, the organization can create a sense of urgency and accountability, reducing the chance of lingering non-compliance.

Address Process At Every Level; Identifying and addressing each reporting level within the organization's pyramid structure is crucial. By understanding unique reporting requirements and ensuring compliance throughout the hierarchy, the organization can avoid cracks in the system and set a solid foundation for success.

What Should Have Been Done Differently?;

To avoid the pitfalls experienced in this incident, the organization could have taken the following actions:

Robust Change Management Plan; Develop a comprehensive change management plan that includes strategies for securing buy-in, communicating the transition, and addressing non-compliance at every level of the organization.

Training and Support; Provide adequate training and support to all employees involved in the CRM implementation. Address any concerns or challenges they may face in transitioning from old systems to the new CRM, ensuring they have the tools and knowledge to embrace the change fully.

Monitoring and Accountability; Implement monitoring mechanisms to ensure compliance with the new system. Regularly assess the use of old systems and provide feedback and consequences for non-compliance. This helps reinforce the importance of adhering to the new CRM and fosters a culture of accountability.

BY PROACTIVELY ADDRESSING NON-COMPLIANCE, effectively managing change, and promoting accountability, organizations can maximize the benefits of a modern CRM, streamline operations, and improve overall sales effectiveness.

19

External User Adoption

When it comes to customer satisfaction, many think you have everything under control if you don't have any complaints and are being polite and addressing small concerns as they pop up. But trust me, making your customers happy goes beyond keeping everything on course and being courteous. It requires you to meet and exceed their expectations with every interaction. Let me emphasize that leveraging the external-facing functionalities of CRMs is essential to achieve optimal customer satisfaction.

One crucial aspect of CRMs that requires advanced understanding is personalization. Personalization is more than addressing your customer by name. It's about anticipating their needs and preferences before they even vocalize them, proactively resolving their issues, and creating a tailored customer journey that matches their unique needs. Using the data and insights in a CRM, you can create personalized experiences that delight customers and foster loyalty.

Another advanced topic to consider when leveraging a CRM is taking an omnichannel approach. Customers today expect to interact with businesses across multiple channels, whether email, social media, or phone. By optimizing your CRM to track and manage communication across a variety of channels, you can create a seamless, integrated customer journey regardless of the channel used. This ensures that your customer receives the same high-quality experience across all the touch-points in their journey.

I've been in your shoes (and they hurt), thinking that a small number of dissatisfied customers won't hurt business much. But even a single unhappy customer can spread negative word-of-mouth, harming your organization's reputation and significantly impacting your bottom line. The cost of replacing lost customers is much higher than retaining existing ones, making it crucial to prioritize customer satisfaction.

Leveraging the external-facing functionalities of these systems is essential to enhance customer satisfaction further. Implement personalization, take an omnichannel approach, and always keep customer satisfaction in mind. By doing so, you'll create a positive impact on your business, your employees, and your customers. Feel free to share your experiences to collaborate and learn from each other.

Let's explore key topics related to using CRMs to make external customers happy, offering ideas for addressing and improving each area and potential positive and negative outcomes if these areas are enhanced or neglected.

Personalization

This is crucial to customer satisfaction, demonstrating that an organization understands and values its customers. CRMs can collect and analyze customer data, enabling businesses to tailor their marketing efforts, product recommendations, and customer interactions based on individual preferences and behaviors.

Ideas for improvement:

- Use CRM data to **segment customers** based on demographics, purchase history, and browsing behavior, allowing targeted marketing campaigns and personalized offers.

- Implement **personalized product recommendations** on the company website or email marketing campaigns, using CRM data to determine which products will most likely appeal to each customer.
- **Train customer service representatives** to use CRM data to personalize their interactions with customers, addressing them by name and referencing past purchases or inquiries when appropriate.

Potential outcomes:

- **The Good:** Personalization can increase customer satisfaction, loyalty, and repeat business, ultimately boosting revenue and brand reputation.
- **The Bad:** Neglecting personalization can result in generic, impersonal customer interactions, leading to decreased customer satisfaction and loss of business.
- **The Ugly:** Over-personalization can result in increased customer support tickets and increased maintenance - thus further spreading the support team further

Customer Communication

Effective communication is essential for maintaining customer satisfaction, and CRMs can streamline and enhance customer communication efforts. By leveraging CRM functionality, businesses can ensure customers receive timely, relevant, and consistent communication across all channels.

Ideas for improvement:

- Use CRMs to automate email marketing campaigns, ensuring customers receive regular updates, promotions, and personalized offers.
- Implement a CRM-integrated live chat or chatbot feature on the company website, allowing customers to receive immediate assistance with their inquiries.

- Utilize CRM data to identify customer preferences for communication channels (e.g., email, phone, or social media) and tailor communication efforts accordingly.

Potential outcomes:

- **The Good:** Effective communication can increase customer satisfaction, loyalty, and positive word-of-mouth, ultimately boosting revenue and brand reputation.
- **The Bad:** Neglecting customer communication can result in dissatisfied customers, loss of business, and damage to the organization's reputation.
- **The Ugly:** High levels of communication may require additional resources, and custom (non-automated) communications will need the appropriate level of supervision, or the risk of offending becomes high

Customer Support

Providing exceptional customer support is a critical factor in maintaining customer satisfaction, and CRMs can be utilized to improve the efficiency and usefulness of support efforts. By leveraging CRM functionality, businesses can ensure that customer inquiries are addressed promptly and that support representatives can access the necessary information to resolve issues effectively.

Ideas for improvement:

- Integrate CRMs with customer support platforms, such as helpdesk software or live chat tools, allowing support representatives to access customer data and history during interactions.
- Use CRM data to identify common customer issues and develop resources, such as FAQs or knowledge base articles, to address these concerns proactively.

- Implement a ticketing system within the CRM to track and manage customer support inquiries, ensuring that issues are resolved on time and that no questions fall through the cracks.

Potential outcomes:

- **The Good:** Improved customer support can increase customer satisfaction, loyalty, and positive word-of-mouth, ultimately boosting revenue and brand reputation.
- **The Bad:** Neglecting customer support can result in dissatisfied customers, loss of business, and damage to the organization's reputation.
- **The Ugly:** As stated earlier, increased staff may be required, and deficient support may cause bad "reviews" and comments on social media.

Customer Feedback and Continuous Improvement

Actively seeking and utilizing customer feedback is essential for continuous improvement and customer satisfaction. CRMs can collect, analyze, and act on customer feedback, helping businesses identify areas for improvement and implement changes based on customer input.

Ideas for improvement:

- Use CRM-integrated survey tools or feedback forms to collect feedback on various aspects of the business, such as product quality, customer service, or overall satisfaction.
- Analyze customer feedback data within the CRM to identify trends, patterns, and areas for improvement.
- Establish a cross-functional team responsible for reviewing feedback and making changes based on the

insights gathered, ensuring customer input is considered in decision-making processes.

Potential outcomes:

- **The Good:** Actively seeking and utilizing customer feedback can lead to continuous improvement, increased customer satisfaction, and a more decisive competitive advantage.
- **The Bad:** Ignoring or dismissing customer feedback can result in stagnation, dissatisfied customers, and loss of business.
- **The Ugly:** Things mentioned in "The Bad" can spiral quickly out of control and become viral. This is usually far faster than a company can react to remedy.

Customer Retention and Loyalty

Maintaining customer retention and loyalty is crucial for long-term business success, and CRMs can be used to monitor and improve these metrics. Businesses can leverage CRM functionality to identify at-risk customers, develop targeted retention strategies, and reward loyal customers.

Ideas for improvement:

- Use CRM data to identify customers who may be at risk of churning, such as those with declining purchase frequency or engagement, and develop targeted retention campaigns to re-engage these customers.
- Implement a CRM-integrated loyalty program, offering rewards or incentives to customers based on their purchase history and engagement with the brand.
- Leverage CRM data to identify high-value customers and develop personalized offers or experiences to strengthen their relationship with the brand.

Potential outcomes:

- **The Good:** Improved customer retention and loyalty can lead to increased revenue, reduced customer acquisition costs, and a more decisive competitive advantage.
- **The Bad:** Neglecting customer retention and loyalty can result in increased churn, loss of revenue, and damage to the organization's reputation.
- **The Ugly:** Dissatisfied customers that use an online system can quickly become a "cancer" and infect others with their dissatisfaction and can also direct people to your competition.

MAKING external customers happy is paramount for any organization, and leveraging the externally facing functionality of CRMs can play a large role in achieving this goal. Businesses can enhance customer satisfaction and drive long-term success by focusing on personalization, customer communication, customer support, customer feedback, customer retention, and loyalty. However, neglecting these critical areas can lead to dissatisfied customers, loss of business, and damage to the organization's reputation. By continually striving to improve these aspects of their operations through CRMs, organizations can ensure that they remain competitive, profitable, and customer-centric in today's ever-evolving business landscape.

20

The System Administrator - Core Traits For CRM

The Perils Of Having An Incompetent Administrator

You need to be aware of a few things when using a CRM.

One of them is that your administrator can do massive damage by making seemingly minor changes. If they are too junior, ill-trained, have limited insight/understanding into business processes, or don't give a crap, even if they're just migrating a field around on the screen, getting it wrong can cause havoc with users and business procedures they weren't even aware of.

It's a careful balance to work without a devoted administrator. It would be best to weigh the cost of a dedicated solely administrator against the CRM investment you are making. You must also consider how much you've decided to invest in your current employees and the other roles they're filling.

In the best-case scenario, they can complete both tasks and keep your CRM operational. You can consider hiring an administrator and have more qualified and/or senior staff train them when the system outgrows their strengths. This can work well if you manage your expansion and hire a dedicated administrator before the CRM becomes daunting.

If you think your CRM will save you from a lack of an administrator, think again.

Your CRM will only do what you tell it to do—and if your admin is inexperienced or incompetent, they can cause a lot of damage by making seemingly minor changes. Even if they're migrating a field around on the screen, getting it wrong can cause havoc with users and business procedures they weren't even aware of.

It's a careful balance to work without a devoted administrator. It would be best to weigh the cost of a singularly dedicated administrator against your CRM investment. You must also consider how much you've decided to invest in your current employees and other roles they're filling. In the best-case scenario, they can complete both tasks and keep your CRM operational. You can consider hiring an administrator and have your staff train them when the system outgrows their strengths.

What if I told you that your administrator could single-handedly destroy your entire business?

Yeah, it sounds ridiculous. But think about it: even a minor change can significantly impact your operations. And if you're working with an administrator who's not up to snuff, they might make one of these mistakes—or worse!

So how do you keep yourself safe? How do you ensure that your business isn't come crashing down around your ears because of some simple mistake? Well, first things first: hire the best administrator possible. That way, even if they make a mistake, it won't be so bad. And hey—you'll also be able to keep them around longer and invest in training them!

But what if that doesn't work out? What if your administrator has been making mistakes for years, and there's no way of fixing them? Well, then, maybe it's time for a new approach.

Must Have's For Administrators

LET'S dive into the must-have traits for a System Administrator in CRM and the associated risks if these traits are lacking.

Technical Expertise: A System Administrator should deeply understand the CRM and its underlying technology. They should be proficient in system configuration, customization, and integration. Without this expertise, the administrator may struggle with troubleshooting, optimizing performance, and leveraging advanced features.

RISKS **Of Not Having Adequate Level Include:**

- Inefficient system setup and configuration.
- Inability to address technical issues promptly, leading to system downtime.
- Missed opportunities to leverage advanced functionality for improved user experience and business outcomes.
- **Problem-Solving Skills**: Being a System Administrator often means encountering complex problems and finding practical solutions. A skilled administrator can analyze issues, identify root causes, and implement appropriate fixes. Without strong problem-solving skills, the administrator may struggle to overcome challenges and resolve system-related issues.

Risks Of Not Having Adequate Level Include:

- Lengthy system downtime due to unresolved issues.
- User frustration and decreased productivity caused by recurring problems.
- Inability to adapt to changing business requirements and address new challenges effectively.

- **Attention to Detail**: CRMs handle vast amounts of sensitive customer data. An Administrator must pay close attention to detail to ensure data accuracy, integrity, and security. Without this trait, the administrator may overlook critical details, resulting in data inconsistencies, privacy breaches, and compliance violations.

Risks Of Not Having Adequate Level Include:

- Data inaccuracies lead to poor decision-making and customer dissatisfaction.
- Security vulnerabilities and data breaches due to weak data protection measures.
- Non-compliance with data privacy regulations, resulting in legal and financial consequences.
- **Communication and Collaboration**: Effective communication skills are vital for System Administrators to interact with users, stakeholders, and technical teams. They should be capable of translating technical jargon into clear, understandable language. Without strong communication and collaboration abilities, the administrator may struggle to gather requirements, convey system updates, and drive adoption.

Risks Of Not Having Adequate Level Include:

- Miscommunication leads to misunderstandings, delays, and confusion.
- Resistance to system changes and low user adoption rates.
- Difficulty coordinating with other teams, resulting in inefficient processes and missed integration opportunities.
- **Continuous Learning**: Technology advances rapidly, and a proficient System Administrator must be willing to learn and adapt. Staying updated with industry trends,

attending training sessions, and exploring new features and functionalities are crucial. Without a thirst for continuous learning, the administrator may fall behind on system updates and miss opportunities to improve processes.

Risks Of Not Having Adequate Level Include:

- Outdated system configurations that inhibit efficiency and performance.
- Limited knowledge of new features and functionalities, reducing the ability to optimize the system.
- Inability to provide adequate support and training to users.

Investing in a System Administrator who possesses these traits can make a significant difference in the success of your CRM implementation. Recognizing the risks of lacking these traits allows you to prioritize the necessary skills and qualities when selecting and nurturing your CRM team.

Keeping the Risks at Bay: When Good Traits Go Bad

SOMETIMES, too much of a good thing can also be an issue. For example, the following issues have been unfortunately repeated far too often.

Technical Know-It-All; Having extensive technical knowledge is undoubtedly valuable for a System Administrator. However, if they become too focused on technical details, they might lose sight of the bigger picture: the user experience. Without balancing their tech focus with a user-centric perspective, the administrator may overlook the needs and preferences of the end-users. Cons associated with this trait include:

Disconnected user experience; The CRM may be optimized from a technical standpoint but fail to meet user expectations, resulting in low adoption rates and decreased productivity.

Lack of empathy; The administrator may struggle to understand user pain points and challenges, hindering their ability to address user needs and provide adequate support.

Supreme Solitaire; Superior problem-solving skills are crucial for a System Administrator. However, isolating themselves and maintaining a solitary approach to problem-solving can harm the success of the CRM implementation. The cons of this trait include the following:

Missed collaboration opportunities; Solitary problem-solving may prevent the administrator from seeking input and insights from others, resulting in suboptimal solutions.

Lack of diverse perspectives; Without involving the expertise and experience of other team members, the solutions may be limited and not fully aligned with the organization's goals.

Tech-Talk Turmoil; Communication skills are essential for a System Administrator, but effective communication goes beyond speaking technical jargon fluently. If the administrator fails to decipher technical jargon and communicate clearly and concisely, it can lead to confusion and misunderstandings. The cons of this trait include the following:

Miscommunication; Inaccurate or misinterpreted information may lead to errors, delays, and decreased efficiency.

Exclusion Of Non-Technical Stakeholders; If technical jargon is not translated correctly, it alienates non-technical stakeholders and inhibits effective collaboration and decision-making.

Leader Or Dictator; Strong leadership skills are valuable, but it is vital to strike the right balance. If a System Administrator becomes overly authoritarian and employs an iron fist approach, it can harm team morale and collaboration. The cons of this trait include the following:

Reduced motivation and engagement; A dictatorial leadership style can stifle creativity, autonomy, and job satisfaction, leading to demotivated team members.

Lack of open communication; Team members may hesitate to share ideas, feedback, and concerns, leading to missed opportunities and increased turnover.

By recognizing these potential pitfalls, you can guide your System Administrator toward a balanced approach that fosters collaboration, effective communication, and a user-centric perspective, ultimately increasing the success of your CRM implementation.

Hiring or Handpicking Staff: The Groundbreaking Decisions

LET'S dive into the points to consider when identifying new staff or promoting from within and the cons if your system administrator doesn't have these traits:

From Within: When looking to promote from within, it's essential to identify potential candidates and assess their learning capacity. Investing in training for specialized skills can give your in-house talent the push they need to excel.

However, the cons of promoting from within without considering these traits include the following:

- **Limited perspective:** Without bringing in fresh ideas and perspectives, your CRM team may miss out on innovative solutions and approaches to challenges.
- **Lack of diversity:** Promoting from within can result in a team with similar skill sets and backgrounds, limiting the range of expertise and experiences.
- **New Hires:** Hiring new staff brings fresh perspectives and diverse skill sets. However, it's crucial to carefully weigh the benefits against the costs, training time, and cultural fit. The cons of hiring new staff without considering these factors include:
- **Long onboarding period:** Training new hires takes time and resources, which can delay implementation and impact productivity.
- **Cultural alignment issues:** If new hires do not align with your team's culture, it can lead to conflicts, low morale, and a lack of collaboration.
- **Culture Clash:** Hiring someone not in sync with your team's culture can create a dramatic and disruptive environment.

Cons associated with a culture clash include:

- **Decreased team cohesion:** Misalignment in values and working styles can lead to conflicts, low morale, and reduced collaboration.
- **Negative impact on productivity:** When team members are not on the same page and do not work well together, it can hinder productivity and the successful implementation of your CRM.
- **The Know-it-Alls:** Overconfidence in a system administrator can be detrimental. When someone believes they have all the answers, they may overlook important details and make critical mistakes. The cons of having a know-it-all administrator include the following:
- Lack of receptiveness to feedback: A know-it-all attitude can prevent the administrator from being open to suggestions or considering alternative solutions.
- **Increased risk of errors:** Overconfidence can lead to oversight and the failure to thoroughly test and validate CRM configurations, potentially resulting in system malfunctions or data inaccuracies.
- **The Skill Shortcoming:** Ensuring your system administrator's skill level aligns with the role's requirements is crucial. Hiring over-qualified or under-qualified individuals can both pose challenges. The cons associated with skill shortcomings include the following:
- **Over-qualified hires:** Over-qualified individuals may feel unchallenged and disengaged, leading to a lack of motivation and potential turnover.
- **Under-qualified hires:** Hiring individuals who lack the necessary skills and experience can result in subpar performance, inefficiencies, and increased training needs.

When preparing job postings and hiring decisions, remember that it's a marathon, not a sprint. Take your time and consider these factors:

- **Job Descriptions:** Craft job descriptions that communicate the required skills and qualifications. Avoid exaggeration and be transparent about must-haves and nice-to-haves to attract suitable candidates.
- **Skill vs. Diva-ness:** While skills are crucial for a system administrator, don't overlook attitude and teamwork. Seek a balance and prioritize candidates with a positive attitude and strong interpersonal skills.
- **Interview Process:** Make the interview process a bridge to assess the applicant's technical skills, problem-solving abilities, communication, and team skills. Design thoughtful interview questions and scenarios to reveal a candidate's potential fit for the role and team.

Note: In particular, when selecting a system administrator, choose wisely, invest in training, and remember that adaptability is critical. Nobody is perfect, but by considering these traits and conducting a thorough selection process, you can increase the likelihood of an effective CRM software implementation that meets your and your employees' needs.

Adaptability

THE IMPORTANCE of hiring a highly adaptive system administrator cannot be overstated. As a seasoned professional in the field, I have witnessed firsthand the positive impact that adaptability brings to an organization. Let me share the benefits and potential challenges of having a highly adaptive system administrator, offering you guidance and cautionary advice based on my years of experience.

Having a highly adaptive system administrator brings numerous benefits to an organization. Firstly, adaptability enables the system administrator to respond quickly to technological advancements and changes in business requirements. Inevitably, technology evolves rapidly, and business needs are constantly shifting. By hiring a highly adaptive system administrator, the organization ensures that its systems remain up-to-date and synchronized with the evolving needs of the business. This agility saves time and resources, allowing the organization to stay on the curve with technological innovation and maintain a competitive edge.

Secondly, a highly adaptive system administrator possesses strong problem-solving skills. This is invaluable in a field where challenges and issues arise regularly. An adaptive administrator can analyze, diagnose, and troubleshoot problems efficiently, exploring alternative approaches to find creative solutions. Their ability to adapt to different scenarios and think outside the box ensures that operations run smoothly and disruptions to the business are minimized. Effective problem-solving enhances the efficiency and effectiveness of the organization as a whole.

However, it's important to acknowledge potential pitfalls when hiring a highly adaptive system administrator. One challenge that arises is the risk of information overload. The technology landscape constantly evolves, presenting a constant stream of new tools, techniques, and systems. A highly adaptive system administrator must continuously stay updated and acquire new knowledge to remain effective. However, this continuous learning can lead to information overload and the pressure to keep up with an expanding knowledge base. The system administrator must balance staying informed and preventing burnout from excessive information consumption.

Another potential drawback of adaptability is the possible lack of specialization. Highly adaptive system administrators are often generalists, capable of handling diverse tasks and responsibilities. While versatility is beneficial in many situations, there may be instances where specialized expertise is essential. Depending on the organization's needs, specific technologies or systems may require deep knowledge and technical skills. Hiring a system administrator who is highly adaptive should be balanced with ensuring the acquisition of specialized knowledge where it is most relevant to the organization's operations.

Hiring and having a highly adaptive system administrator is paramount in today's ever-changing technology landscape. The ability to respond quickly to change and strong problem-solving skills bring immense value to an organization. However, it's crucial to be mindful of potential challenges such as information overload and the need for specialization in certain areas. Hiring a highly adaptive system administrator should be approached with due diligence, ensuring that the individual possesses a balance of adaptability and the necessary expertise that aligns with the organization's specific requirements. By embracing adaptability while maintaining a strategic focus on specialization, organizations can maximize their potential for success and ensure the effective management of their technological systems.

21

Case Study: Unapproved Outsourcing

Scene:

Imagine when the modern workplace had transitioned to embrace remote work and telecommuting, becoming integral to the everyday fabric of work life. In this bustling office, employees were no longer confined to traditional cubicles and conference rooms. Instead, they were scattered throughout the room, each with their laptops open and headphones on, creating an atmosphere of focused productivity.

As you enter this dynamic environment, you see the seamless integration of technology into the workday. Video meetings took place, with employees engaging in virtual conversations with colleagues from near and far. The occasional ping of a direct message could be heard, indicating quick and efficient communication between team members across various locations.

The scene is one of collaboration and independence, with

employees embracing the freedom and flexibility of remote work. Technology enabled these individuals to work together seamlessly despite the physical distance, ensuring that projects progressed smoothly and deadlines were met.

The CRM would serve as the backbone of the business, enabling employees to access critical data, collaborate on projects, and stay connected regardless of location. In this evolving landscape, companies had to ensure that their chosen CRM was robust, reliable, and adaptable to remote work demands. It was imperative to select a system that offered seamless integration with remote tools, enabling employees to access customer information, manage tasks, and track progress from any location. Furthermore, the CRM had to be user-friendly and intuitive, easing the transition for employees from the traditional office setup to the remote work environment.

Incident:

As telecommuting became increasingly common, so did a report of an individual outsourcing their work. He faithfully commuted daily to the physical office, diligently completing his projects. His work product was nothing short of exemplary. However, his manager began to notice a peculiar discrepancy. While the employee excelled at communicating technical matters through email, he struggled to convey the same level of understanding in person. Something was amiss.

Recognizing the need for further investigation, the manager quietly researched the situation. She started by examining the system logs, hoping to gain insight into the employee's activities. What they uncovered was genuinely perplexing. It became evident that the worker had been logging in remotely during the late-night hours, engaging in a whirlwind of activity. It was an unfortunate and baffling revelation.

Upon further analysis, it became clear that this individual had been outsourcing his own work to a group of experts in his home country. While he slept, this remote network of skilled professionals was proficiently performing his tasks. The magnitude of the deception was astounding.

This incident serves as a stark reminder of the potential risks that can arise when employees blur the boundaries of their responsibilities. As telecommuting becomes more prevalent, organizations must exercise due diligence in monitoring employee performance, ensuring that the work being attributed to them is, in fact, a true reflection of their efforts.

The company had a vigilant manager who recognized the discrepancy in this case. However, it is a **cautionary tale for all organizations.** It underscores the importance of implementing robust oversight mechanisms, conducting regular performance assessments, and fostering an environment of open communication and accountability. By doing so, businesses can mitigate the risks associated with such deceptive practices and build trust and integrity within their workforce.

Let this incident serve as a reminder of the importance of upholding professional ethics and maintaining vigilant supervision, even in the evolving landscape of telecommuting. By doing so, we can ensure that the work being carried out is truly a product of the efforts invested by the employees while safeguarding the organization's reputation and integrity.

Result:

In this unfortunate incident, the repercussions were far-reaching, causing considerable problems for the company. At the forefront of these issues were the apparent security breaches and ethical violations that had taken place. The decision by the individual to outsource their work to an overseas team, one that contained sensitive and confidential information, immediately posed significant security risks. Such an action was unacceptable and could have disastrous consequences for the organization.

Moreover, the individual's acceptance of full payment for work being carried out by individuals in another country violated basic ethical principles of honesty and integrity. The decision to deceive the company in this way was unacceptable. It had rightfully caused alarms to sound among the executive team.

Given this flagrant breach of trust, the individual's immediate termination was necessary. While unfortunate, it was clear that such a level of deception and misconduct had no place in the organization. The matter was dealt with expeditiously and decisively, in keeping with the company's policies for addressing such severe breaches of conduct.

This warns organizations across all industries that such actions' risk and potential consequences cannot be overstated. The importance of due diligence and close monitoring of employees cannot be understated. Additionally, the company should ensure that proper security protocols and ethical standards are in place to minimize the possibility of such breaches of trust and integrity. Ultimately, the organization and its employees are responsible for upholding the highest standards of ethical conduct and professionalism.

Lessons Learned:

Ensure Clear Guidelines For Remote Work; Companies must clearly define what is and is not allowed. There should be strict rules and policies around outsourcing work and only delegate to trusted sources.

Monitor Remote Workers; Companies must monitor remote workers' activities and work habits. Regular check-ins and meetings help ensure that the work is being done.

Implement Additional Security Measures; Security is crucial when employees are working remotely. Companies should include other security measures, such as IP address blocking and timezone blocking, that limit where and when employees can log into the company's systems.

What Should Have Been Done Differently:

To avoid this situation, companies should consider following these recommendations:

Screen job candidates thoroughly; Companies must thoroughly vet and screen them before hiring them. It is essential to ensure that the candidate is the right fit for the job, both technically and ethically.

Limit access to sensitive information; Companies should limit sensitive information to only those who need it. Access should be granted based on necessity and security clearance.

Perform regular audits and assessments; Companies should conduct regular audits and assessments to detect potential security risks and breaches. This will help to mitigate the risk of security breaches and prevent outsourcing attempts.

Overall, remote work is a valuable benefit that benefits employees and companies alike. However, it is essential to implement appropriate policies and security measures to avoid potential risks while embracing the strengths of the remote work lifestyle.

22

The Battle For Data: Fight The Good Fight

In the world of CRM, sales professionals, and consultants are notorious for their attachment to data. They cherish their data more than kids love candy on Halloween. They might as well put a "Do Not Disturb" sign on their heaps of spreadsheets and folders. However, it's essential to note that these individuals are not necessarily hoarding. Instead, they understand the value of data in today's world, and they recognize its importance to their operations.

However, the potential for future success is enormous if we can get these data enthusiasts to share their knowledge. That said, this is easier said than done. Professionals often dislike sharing their data because they fear losing control, perceived expertise, or job security. Nonetheless, there are ways to mitigate these concerns. By identifying the benefits of collaboration and information sharing and highlighting success stories where shared data led to significant achievements, professionals can be persuaded to change their attitude towards data sharing.

Embarking on the journey to optimize your CRM system can often feel like solving a cryptic puzzle, with hidden possibilities awaiting discovery. However, be prepared for the hurdles that lie ahead. To unlock the true potential of your CRM, the first step is a moment of pause to scrutinize all your data repositories thoroughly. While the task of analyzing data may be arduous, it is vital to identify potential roadblocks that could derail your progress down the line.

One significant challenge you will encounter is making decisions that may not be popular and could raise eyebrows among your colleagues. However, it is essential to understand that these decisions are crucial in establishing a solid foundation for your CRM system. This foundation ensures its long-term efficiency and functionality, setting the stage for success.

But let me be clear; this is not a one-time endeavor. Building and maintaining an effective CRM system requires ongoing analysis and evaluation. System administrators must remain vigilant in identifying potential problems related to data sharing and implement proactive solutions to mitigate these issues. This involves establishing a backbone for your CRM that fosters a transparent and cooperative environment encouraging collaboration and information sharing throughout the organization.

Navigating the complexities of data sharing and cultivating a culture of collaboration may present challenges. However, it is an essential undertaking that guarantees long-term efficiency and functionality in your CRM strategy. One approach to instilling this shift is directly addressing professionals' concerns and openly discussing the advantages of knowledge sharing. By highlighting the value and impact of collaboration, you can gradually change mindsets and foster a culture that enhances your CRM efforts.

Remember, sustaining success requires constant attention, analysis, and evaluation. Continually evolve your CRM system by staying informed of industry trends, emerging technologies, and best practices. Regularly audit your data repositories and make any necessary adjustments to ensure the accuracy and integrity of your CRM system. Adaptability and flexibility are vital qualities in a successful CRM strategy.

By meticulously examining data repositories, making tough decisions, fostering collaboration, and maintaining a constant focus on evaluation and improvement, you can build a strong foundation for your CRM system. Embrace the opportunities that lie ahead and unlock the full potential of your CRM to drive growth, enhance customer relationships, and achieve lasting success.

Data, Being Strict but friendly about it.

NOW, I know data might not be the most exciting subject, but trust me, in the world of CRM, it's pretty damn crucial, and I think it's time the attitude around this changes. The issue here isn't just about having crappy data in the system; **the real problem is the lack of appreciation for how valuable high-quality data can be.**

Think about it this way: If your CRM is running on dirt, it's going to give you shit results. Pardon the language, but it's true. A CRM is only as good as the data you feed into it. When dealing with customer data, we need a level of uncompromising strictness, one that's frequently not understood, respected, or appreciated.

So if protecting data quality is so important, why do I see it neglected repeatedly? The truth is, it often comes down to a lack of education on the matter. People don't realize how crucial it is; unfortunately, it's not a lesson you want to learn the hard way. I mean, I get it. Your customers are busy; they have a million things on their plate, and being a stickler for good data practices might feel out of their comfort zone, but it's worth revamping everyone's approach to data collection, organization, and maintenance. Better to put in the effort now than troubleshoot a mess later.

Many people have built their kingdoms on maintaining non-standard, headache-inducing data. They've mastered the art of turning this messy data into tools, reports, and dashboards – mental gymnastics at best job preserving tasks at worst. These gurus of data gymnastics have become indispensable, the gatekeepers of reliable and actionable information. But here's where we collectively slap ourselves back to reality. Time-consuming, difficult-to-modify data is not something we should be proud of. Sure, these wizards have learned the tricks and traps of the trade, but our mission is to move beyond those chaotic days to a new era of sleek, standardized, high-quality data.

TEACH YOUR TEAM – your wizards and even the paper pushers – that it's time to leave the crumbling castles of non-standard data behind. Explain the incredible power and potential that high-quality data possesses. Keep preaching that gospel until every team member can see the light.

WELL-MEANING TEAMMATES, external consultants, backseat drivers – they all come swarming in with pearls of wisdom, advice, and their darling suggestions. And sure, that's part of the deal. To pile onto the chaos, you've also got a collection of systems, each more stubborn than a mule, with their predetermined data rules and formats. Brace for impact because with shiny new systems come shiny new rules that might initially throw your customers and clients for a loop. Yes, just when they thought they'd gotten comfy with the old data formats, we're shaking up the snow globe. So, let's tackle these changes head-on. Don't shy away or procrastinate because the truth is that these changes are the 'new normal.' Once the dust settles, people *will* acclimate to the new formats and processes.

Remember, the kind of change we're discussing here isn't just change for the sake of it. It's strategic, progressive, and necessary to keep any modern business running like a well-oiled machine.

Proper Path and Consideration

REGARDING DATA CLEAN-UP AND INTEGRATION, a few key factors can significantly impact your journey to success. To ensure that your effort is effective and meets your goals, it's crucial to consider the following points carefully. Let's delve into each of these factors to understand their importance better.

Migrating Data; One of the most obvious aspects to consider is whether or not you should migrate your data. This choice can have an enormous influence on the overall success of your project. Migrating data can be a complex process, requiring careful planning and execution. It's crucial to weigh the benefits and potential challenges before making any decisions. Consider what data needs to be migrated, the systems' compatibility, and the potential impact on your business operations. In some cases, starting fresh with clean data might be a more viable option.

Customer Expectations; Another crucial factor to consider is customer expectations. Ultimately, the success of your data clean-up and integration efforts will be determined by your customers' ability to use the new system effectively. Understanding their needs, preferences, and desired functionalities is essential. Engage with your customers early to gather feedback and ensure the final solution meets their expectations. This will increase customer satisfaction and drive the adoption and usage of the new system.

Project Budget; Time and money are valuable resources for any project. The budget allocated for your data clean-up and integration project is pivotal in determining its success. Adequate funding allows you to allocate resources effectively, hire skilled professionals, and invest in necessary tools and technologies. However, it's important to strike the right balance between budget constraints and project objectives. Consider different cost-saving strategies, prioritize critical tasks, and continuously monitor and manage your budget to ensure it remains aligned with your goals.

Technical Expertise; Lastly, the willingness to use and access technical expertise is a vital factor that can significantly influence the success of your project. Data clean-up and integration projects often require specialized knowledge and skills. It's essential to assess your team's capabilities and identify any existing gaps. Depending on the project's complexity, you may need to consider external assistance, such as hiring consultants or partnering with a data integration service provider. Access to technical expertise can expedite the process, mitigate risks, and improve the overall quality of your data.

BY CAREFULLY CONSIDERING these four factors, you can set yourself up for success in your data clean-up and integration journey. Each aspect plays a crucial role in determining the outcome of your efforts. Always weigh the pros and cons of data migration, actively engage with your customers, manage your budget effectively, and seek technical expertise when needed. By doing so, you'll be well on your way to achieving your goals and establishing a solid foundation for your data-driven initiatives.

Time and Money

UNDERSTAND that data clean-up and migration efforts are often complex and time-consuming processes. Even for a medium-sized organization, these tasks can take months and years and cost hundreds of thousands of dollars. The worst part? These projects frequently require a lot of people's time, many of whom are not usually asked to participate in tech efforts. It's crucial to involve the right group of people in these endeavors - the "owners" and users of the data in all its forms - since they have high expectations for the new system and may resist making necessary considerations

When you budget for data clean-up and migration, it's essential to consider not only the financial costs but also the total human hours required, travel expenses, third-party expertise, and training and awareness. The awareness aspect is critical and should not be overlooked or trivialized. Educating stakeholders about the project and its impact can save the project from potential derailment and empower them to make informed decisions. For more insights on this topic (and others), I recommend reading the chapter on Setting Expectations. Data clean-up, migration efforts, and customer data management are crucial for any organization that wants to succeed in today's digital age. Be sure to involve the right people in data efforts, consider all costs and potential roadblocks, and focus on efficiently managing and utilizing customer data to improve the customer experience.

23

Battle Royal - The Decision To Migrate Data vs. Going With A Clean Slate

In my years of experience, I have witnessed countless companies making the critical decision of how much data to migrate when transitioning to a new system. This is an issue of grave importance that demands careful consideration. This topic is so significant that it has been thoroughly discussed in our own Battle Royal - The Decision to Migrate Data vs. Going With A Clean Slate. Throughout my career, I have seen many organizations grapple with this decision, often succumbing to the temptation of migrating every last bit of their existing data. However, it is imperative to approach this decision with wisdom and caution, as the consequences of an uninformed choice can be substantial.

I have observed firsthand the aftermath of companies that believed it necessary to migrate vast volumes of existing data, only to regret that decision later. This regret becomes apparent through the arduous pains experienced during the data clean-up and integration process and the constant struggle of comparing the old versus new data. Allow me to illustrate this by providing you with an example.

Imagine using an outdated CRM system with inconsistent data formatting and peculiarities. If you were to migrate all of this old data to the new system, the persisting inconsistency would make it exceedingly challenging to compare and effectively utilize the new data format. The presence of outdated data may also foster a longing for the old ways, leading to an attachment to old reports and established summaries. However, it is crucial to shift focus towards creating new and improved practices that align with the capabilities and advantages offered by the new system.

Additionally, it is essential to recognize that deciding how much data to migrate extends beyond the technical aspects. It involves considering the existing data's usability, relevance, and integrity. It requires a thorough evaluation of the specific data elements that are vital for your CRM efforts' ongoing success. By carefully analyzing the value and impact of each data component, you can make informed decisions that drive efficiency and support strategic decision-making.

As you contemplate the decision of data migration, it is paramount to exercise due diligence and seek guidance from seasoned professionals. Engaging in thorough research and consulting with experts in the field will provide valuable insights and help you navigate the potential pitfalls associated with this decision. Furthermore, conducting pilot tests and gradually migrating data in stages can serve as a prudent approach to mitigate risks and validate the effectiveness of the chosen migration strategy.

The decision of how much data to migrate when transitioning to a new CRM system should not be taken lightly. While the temptation to migrate every last bit of existing data may be substantial, it is essential to approach this decision with wisdom and caution. Understanding the potential consequences and aligning your choices with your organization's overarching goals and needs is paramount. By making informed decisions and seeking expert guidance, you can ensure that your data migration efforts are driven by purpose and contribute to the success of your CRM initiatives.

. . .

TO AVOID THESE PITFALLS, it's crucial to approach data migration pragmatically, focusing on the bigger picture. Consider the following factors:

Evaluate the quality of your existing data; Take a close look at the quality of your data and determine whether it's worth migrating. If you've been using an outdated system with inconsistent data formatting and various other issues, it might be best to start from scratch and only migrate the essential data.

Consider the importance of the data; Determine which data is crucial to your organization's operations and start by prioritizing the migration of that data. By focusing on the most critical data, you can avoid the regret of migrating too much data unnecessarily.

Map out your new system; Plan how the new system will be structured and how the data will integrate. Consider what changes and improvements you would like to make. This planning stage will give you a clear idea of what data you need to migrate.

Decide on the timeline; Establish a timeline for data migration and ensure that all stakeholders are aware of the timeline. This will help to manage expectations and ensure that the process is completed in a timely fashion.

BY FOLLOWING THESE STEPS, you can approach data migration with confidence and avoid the pain of comparison between old and new data formats. It's all about taking a pragmatic approach and focusing on what's essential to your organization's success.

24

Clean Slate

In data migration, deciding to "wipe the slate clean" and start afresh may initially appear extreme. However, I cannot stress enough how this approach can often prove to be the right choice, providing many benefits that minimize headaches, stress, and costs, if available.

At a high level, the Clean Slate method champions the notion of a rebirth, a fresh start filled with infinite possibilities. It enables you to reverse any previous decisions that may have skewed the playing field in the old system, favoring the needs of a select few at the expense of others. The prospect of a new system unburdened by the baggage of the past can work wonders, even bordering on magic, if it aligns with your organization's specific needs and circumstances. Moreover, introducing a new system without migrating old data removes the element of comparison, erasing objections that invariably arise when transferring outdated data. A clean slate motivates individuals to wholeheartedly embrace the new system and its processes, viewing it as an opportunity to work with fresh, accurate data and explore uncharted possibilities.

One of the key advantages of the Clean Slate approach is the freedom it bestows on newer and smaller teams. Freed from the limitations and complexities imposed by legacy data, they can devote their energies to designing a system that perfectly aligns with their unique requirements. Armed with fresh, relevant data, it's as if they possess the power to mold their destinies without being fettered by the constraints of the past.

Allow me to share a personal anecdote from my experience that exemplifies the transformative power of the Clean Slate approach. Years ago, I assisted a company needing a CRM overhaul. Outdated and inaccurate data plagued the existing system, impeding their ability to make informed decisions. After careful consideration, the company embarked on the Clean Slate path after carefully considering whether to invest time, effort, and resources into migrating flawed data or starting anew. They viewed this as a golden opportunity to shed the limitations of the past and construct a system that genuinely met their evolving needs.

The outcome was nothing short of remarkable. The team embraced the new system with unwavering enthusiasm, unburdened by previous data inconsistencies. This newfound freedom allowed them to envision their workflows in a fresh light, unencumbered by past limitations. With access to accurate and up-to-date data, they were able to unearth previously untapped insights and propel the organization forward. The transformation was nothing short of magnificent, and it stands as a testament to the inherent power of starting anew with a clean slate.

Acknowledging that the decision to embark on a Clean Slate approach should not be taken lightly is crucial. While it holds immense potential, it also warrants thorough consideration and due diligence. The repercussions of such a decision ripple across all levels of the organization, impacting users, customers, the project team, and administrators alike. It is essential to involve all stakeholders, gather their perspectives, and assess your unique circumstances' potential benefits and drawbacks.

The Clean Slate approach to data migration merits careful consideration and assessment. It offers unparalleled opportunities for organizations to break free from the shackles of the past, revitalize their processes, and harness the power of fresh, accurate data. However, it is crucial to engage in diligent evaluation, ensuring alignment with organizational goals and the needs of all stakeholders. By embracing the Clean Slate approach with caution and skill, organizations can embark on a journey of transformation, innovation, and renewed success.

HOWEVER, it's essential to exercise caution when embarking on a Clean Slate journey. Consider the following key points:

Data Evaluation; Before deciding, thoroughly evaluate the quality and relevance of your existing data. Identify any critical information or insights that you may lose by not migrating. Consult with key stakeholders to ensure you're making an informed choice.

Regulatory And Legal Requirements; Assess any regulatory or legal obligations that require you to retain specific data. Compliance should be a top priority, so adhere to any necessary retention periods or protocols.

Stakeholder Engagement; Involve relevant stakeholders in the decision-making process. Seek their input, address their concerns, and communicate the benefits of the Clean Slate approach. By fostering a sense of ownership and understanding, you'll cultivate an environment of support and cooperation.

Data Migration Strategy; If you opt for a Clean Slate, develop a comprehensive strategy for migrating any critical data that genuinely has value in the new system. Ensure a smooth transition by carefully planning and implementing the necessary steps to transfer essential information.

THE CHOICE TO start fresh with a Clean Slate and leave historical data behind can be a game-changer. It's an opportunity to unleash creativity, embrace change, and build a system that meets your unique needs. However, it's essential to approach this decision carefully, considering factors such as data evaluation, regulatory requirements, stakeholder engagement, and a well-thought-out migration strategy. Doing so paves the way for a successful CRM transformation that propels your organization into a new era of growth and success.

Hopefully, the soundest answer is also the most common, that the existing historical data currently resides in a place that can continue to provide the results required.

When undertaking a clean slate approach to migrating data, it is vital to establish data rules and verification efforts right from the start. By doing so, you can ensure the integrity and health of your data. The new system will have the advantage of collecting historical information in the proper format and with a focus on more current relevance. This means that the data will be more useful and up-to-date, providing better insights into your business operations.

One of the benefits of starting fresh with new data is that there is no need to integrate with any third-party systems. This eliminates the complexity and potential issues associated with integrating legacy data. Instead, your integration efforts can be dedicated to exploring and implementing new and improved ways of doing things. This allows for innovative approaches and avoids being constrained by the limitations of the old system.

By not using old corporate data, you eliminate the possibility of encountering problematic "Data Ghosts." These are issues that can arise when trying to cleanse and reconcile outdated or incorrect data from the legacy system. With a clean slate, you can avoid these challenges altogether.

However, it's important to note that access to existing information from the pre-new system will require a hybrid approach. This means that you may need to find a way to access and incorporate relevant data from the old system into the new one. This can present its own set of challenges, but with careful planning and execution, it can be accomplished.

It's worth mentioning that seeking approval and acceptance from your customer or client base may be more challenging when starting with new data. The level of frustration associated with the existing system can greatly impact their willingness to embrace the change. It's crucial to communicate the benefits and improvements that the new system offers to win their support.

A clean slate approach to migrating data offers numerous advantages, such as ensuring data health, focusing on new ways of doing things, and avoiding the pitfalls of legacy data. However, careful consideration should be given to the access and integration of existing information, as well as managing customer sentiment during the transition. With proper planning and a diligent approach, a successful migration can be achieved.

25

Case Study: Full Of Regrets

Scene 1: Planning the Data Migration

A company and its project team embarked on the ambitious task of migrating data from several legacy systems and smaller departmental databases. These systems housed various data formats, completeness levels, and processes for gathering and maintaining clean data. This diverse collection included global firm-wide financial systems, Outlook/Exchange address books, business unit-hosted marketing lists, and even individuals' "little black books" of contacts and business relationships throughout their careers.

Incident:

At first glance, the fastest plan would prioritize easily transferable data due to accepted commonalities. However, this approach would have been detrimental as it would ignore other data sets that required more extensive handling. It was tempting to side with individuals who emphasized the value and effort they had put into nurturing their datasets to their current state.

It may seem that the fastest plan to use this data in a common and singular format and repository would be to simply include what was easily transferable due to accepted commonalities, but would be easy to ignore and very detrimental to do so would be to ignore the other data sets that would require greater massaging. The point being any data that gets ignored to expedite the greater data cleanup cause also allows for that data and its associated processes to continue to grow outside the larger CRM effort. This creates issues down the road and reduces the effectiveness of the larger managed set of data.

Result:

The project team decided to include all the data in a standardized format and repository, underestimating the potential consequences of prioritizing speed over thoroughness. They did not consider that by ignoring certain datasets to expedite the primary data clean-up, those datasets and their associated processes would continue to grow outside the scope of the CRM effort. This shortsighted decision created future issues and reduced the effectiveness of the overall managed data set.

Lessons Learned:

Include all relevant data; When undertaking a data migration project, it is crucial to carefully evaluate and include all relevant data, even if it may initially seem less critical to specific stakeholders. Ignoring certain datasets can lead to unanticipated issues and hinder the overall system's effectiveness.

Prioritize thoroughness over speed; While it may be tempting to prioritize speed and focus only on easily transferable data, such an approach can lead to significant problems and compromises in the long run. Take the time to address all data sets, even if they require more effort to clean and migrate.

Consider the broader impact; Understand that migrating data is not just about technical factors but also the impact on existing processes and workflows. Ignoring certain datasets can lead to disjointed processes and hinder the adoption of the new CRM.

What Should Have Been Done Differently:

In hindsight, it would have been wiser to prioritize a comprehensive approach to data migration. The project team should have taken the time to properly assess and include all relevant datasets, regardless of individual stakeholders' perceived value. This would have involved allocating sufficient resources and effort to clean and integrate all data sets, ensuring a more effective and seamless transition to the new CRM.

26

Determine What CRM Features You Need, And Their Financial Impact

As businesses seek competitive advantages and more efficient customer service solutions, many turn to CRM systems to optimize their interactions with clients. Implementing a CRM system can yield numerous operational benefits, but a significant budget commitment must be thoroughly evaluated. Before investing in a CRM system, executives should conduct an in-depth analysis of both costs and benefits and consider the potential impact of CRM on different aspects of their organization. A comprehensive cost-benefit analysis will reveal whether the investment in a CRM system is worthwhile and what features are necessary to maximize its benefits.

Besides providing an overview of the financial implications of a CRM implementation, executives should consider the system's overall functionality from the outset. A CRM evaluation can help businesses assess and understand how they interact with their clients; it can reveal new ideas and insights into client behavior and preferences. A CRM implementation should be more than just a technical upgrade, as it can offer valuable insights and a bird's-eye view of how the organization interacts with its customers.

The first step in a CRM's implementation process is to evaluate the available features and determine which ones are necessary. However, it is crucial to remember that not all elements are essential for all organizations, and prioritizing which features to implement should be based on business needs. Functionality vs. expense is a crucial consideration: while most CRM systems come with a basic set of functions, others can provide a library of integrations and features that enhance interactions between companies and their clients. One should consider the business' operational requirements, users' need for integration with other systems, and the system's capabilities to decide on the necessary features.

Flexibility is another crucial aspect of a CRM system. It is essential to ensure that the product you choose can provide the flexibility you require, not just at present but also in the future. As companies expand, they will encounter new requirements that necessitate modifications to their CRM systems. One way to address flexibility is to ensure that the CRM system interacts with the various applications that the company uses, enabling employees to access necessary relevant data and provide better services to customers. Customizing the CRM system's features to suit the company's specific needs is an excellent way to improve productivity, eliminate redundancies, and reduce the cost of training and support.

The issue of remote working has gained increasing urgency in the wake of recent pandemics, as companies require remote management capabilities to keep their CRM system functioning efficiently. As working from home has become the new normal for many companies, the ability for the CRM to be accessed from remote locations is vital. However, ensuring that the CRM system is highly secure and that access is granted only to authorized personnel is essential. Cloud-based systems provide flexibility and remote access, minimizing downtime, improving efficiency, and lowering support costs. The best remote CRM solutions offer secure remote access, achieve maximum uptime through efficient hosting and fail-safe systems, reduce support costs, and ensure all security protocols are in line.

Integration capabilities also play a significant role in evaluating a

CRM system. A successful CRM system must integrate with the company's existing applications, databases, and systems, allowing seamless data flow. Therefore, ensuring that the CRM system has a robust API (Application Programming Interface) capable of integrating with other apps and systems is essential. A sound CRM system should efficiently communicate with all client-facing and back-office systems, facilitating the smooth flow of information across the company. Integration can also improve user productivity by allowing real-time access to customer data, interaction tracking, and follow-up statuses via collaboration tools like SharePoint, Skype, Microsoft Teams, or email.

When discussing budgeting, it is crucial to keep the long-term financial impact of the CRM system in mind. Companies often make the mistake of only focusing on the implementation costs, failing to account for the ongoing maintenance and support required to keep the system functioning correctly. To avoid such challenges, one needs to implement a care and support infrastructure that addresses issues quickly and cost-efficiently and provides a high level of support and expertise that helps maximize the CRM system's benefits.

One of the most significant hidden costs associated with a CRM implementation is the cost of customization. Companies may require customization to align the system with their business processes and client interactions. Depending on the degree of customization needed, the cost of implementing a CRM system can spiral quickly, especially when hiring specialized expertise to develop customizations. It is vital to assess how much customization the company requires and then budget accordingly.

A CRM system's implementation is a complex process that requires multiple inputs to be considered before making a sound decision. A comprehensive system cost-benefit analysis based on remote access, integration, customization, flexibility, and functionality, is essential. Business needs should drive the selection of appropriate features to optimize CRM performance. Successful CRM system implementation requires assessing future costs and potential expenses, so the budgeting and planning should account for regular maintenance and support, system upgrades, and ongoing customization when required. Companies that exercise due diligence early in selecting and implementing a CRM system can avoid costly mistakes, enjoy all the advantages of this critical business technology, and position themselves to achieve sustainable growth and success.

27

Resentment Of Cost - Budget Budget Budget

You will encounter many people that will present even more hurdles in implementing and improving the business development process using a CRM.

Undoubtedly, the long-term relationships you build with satisfied clients are the driving force behind any company. Centralizing them is more complex than ever, given every company's fierce rivalry to remain competitive. As a result, investing in integrated CRM software is your best choice for your business.

Convincing leaders that the cost of a CRM is justified becomes much easier when you can demonstrate the **direct impact on the bottom line**. Here are several specific ways to earn more money with a CRM:

Increase Cross-Selling And Upselling Opportunities;_A CRM provides valuable insights into customer behavior, preferences, and purchase history. By leveraging this information, sales teams can identify cross-selling and upselling opportunities, resulting in increased revenue per customer. For instance, if a customer has purchased a laptop, the CRM can suggest related accessories that can be offered to the customer, boosting the average transaction value.

Personalized Recommendations; A CRM can analyze a customer's purchase history and preferences to generate personalized product recommendations. For example, if a customer has bought a camera, the CRM can suggest compatible lenses, tripods, or camera bags. By offering personalized recommendations, businesses can increase the possibility of customers adding more items to their cart, ultimately boosting the average transaction value.

Bundle Offers; A CRM allows businesses to create bundled product offers based on customer purchase patterns. For instance, if a customer has previously purchased a laptop, the CRM can suggest a bundle with a wireless mouse, laptop bag, and antivirus software at a discounted price. These tailored bundles entice customers to upgrade their purchases and increase the overall value of each transaction.

Subscription Upgrades; With a CRM, businesses can track customer subscription plans and identify opportunities to upsell to higher-tier plans. For For example, if a customer is subscribed to a basic streaming service, the CRM can flag them as a potential candidate for an upgraded plan that offers more features or content. This approach can lead to increased recurring revenue and higher customer lifetime value.

Complementary Product Recommendations; A CRM enables businesses to identify products that complement each other and can be offered together. For instance, if a customer has purchased a printer, the CRM can suggest ink cartridges, paper, and maintenance kits as additional items. By highlighting these complementary products, businesses can increase the average order value and enhance the customer's overall experience.

Renewal Reminders; For businesses with subscription-based services, a CRM can automate renewal reminders and proactively alert customers when their subscription is about to expire. This presents an opportunity to upsell customers to a longer-term or premium subscription package. By leveraging the CRM's automation capabilities, businesses can improve renewal rates and generate more revenue from existing customers.

Enhance Customer Retention And Reduce Churn; It's

well-known that getting new customers costs more than retaining existing ones. A CRM allows businesses to work better on improving customer satisfaction and engagement, reducing the likelihood of churn. By providing personalized experiences, tracking customer interactions, and proactively addressing their needs, a CRM helps build better relationships, leading to higher customer loyalty and long-term revenue streams.

Optimize Marketing Campaigns And ROI; With a CRM, marketing teams can gain better insights into customer segments, preferences, and behaviors. Businesses can segment their audience and deliver more targeted and personalized marketing campaigns by analyzing this data. This approach increases the effectiveness of marketing efforts and maximizes return on investment (ROI) by avoiding wasted resources on ineffective marketing strategies.

Segmentation Based On Purchase History; A CRM allows you to segment customers based on purchase behavior. For example, if you run a clothing brand, you can categorize customers who have purchased winter coats in the past. This segmentation enables you to target them with personalized campaigns for new winter arrivals, increasing the chances of driving conversions and sales.

Personalized Email Marketing; With a CRM, you can track customer preferences, interests, and engagement patterns. Leveraging this information, you can create personalized email marketing campaigns tailored to specific customer segments. For instance, if you know a group of customers interested in fitness wear, you can send them targeted emails showcasing new activewear collections or exclusive discounts on fitness accessories. This level of personalization increases engagement and conversion rates.

Remarketing Campaigns; A CRM allows you to identify customers who have shown interest in specific products or services but haven't purchased them yet. With this knowledge, you can run remarketing campaigns targeting these individuals. For example, suppose a customer has added items to their cart but abandoned the purchase. In that case, you can set up automated email reminders or targeted ads offering incentives to complete the purchase. You can boost conversions and maximize your ROI by reminding and enticing hesitant customers.

Real-Time Campaign Tracking; A CRM provides real-time tracking and analytics, allowing you to monitor the performance of your marketing campaigns. You can track click-through rates, conversion rates, and overall campaign ROI. Access to this data allows you to identify which campaigns are effective and allocate resources accordingly. For example, if you notice that paid social media ads drive more conversions than email marketing, you may invest more in social media advertising and optimize your spending.

A/B Testing And Campaign Optimization; With a CRM, you can conduct A/B testing to evaluate the performance of different marketing strategies and tactics. By comparing different versions of your campaigns, such as different subject lines, call-to-action buttons, or visual designs, you can identify what resonates better with your target group and optimize your campaigns accordingly. This leads to improved effectiveness and a higher return on your marketing investments.

Streamline Sales Processes And Shorten The Sales Cycle; A CRM streamlines sales processes by automating tasks, providing centralized access to customer data, and enabling effective collaboration among team members. This enables sales teams to respond faster to customer inquiries, nurture leads more efficiently, and close deals quicker. Shortening the sales cycle allows businesses to generate revenue faster and capitalize on new opportunities.

Automated Lead Management; With a CRM, you can automate lead management processes, such as lead capturing, scoring, and routing. It allows you to automatically assign leads to the appropriate sales representatives based on predefined criteria like territory or expertise. By automating these processes, you ensure that leads are promptly addressed and prevent any delays in response time, resulting in increased opportunities for closing deals faster.

Sales Pipeline Visualization; A CRM visually represents your sales pipeline, giving you a clear overview of where each lead or deal stands in the sales process. This visualization helps identify bottlenecks, prioritize activities, and take necessary actions to move deals forward. With a centralized view of the pipeline, sales teams can focus on deals with a higher chance of closing, shortening the sales cycle and improving overall efficiency.

Task Automation And Reminders; A CRM can automate repetitive tasks and set reminders for sales team members to ensure they follow up with leads and complete crucial actions on time. For example, if a lead hasn't been contacted for a specified period, the CRM can automatically generate a reminder for the salesperson to reach out. These automated tasks and reminders keep the sales team on track, minimizing delays and accelerating the sales cycle.

Collaboration And Communication; A CRM provides a centralized platform for sales teams to collaborate, communicate, and share information. For instance, team members can leave important notes, share insights, or provide updates within the CRM, ensuring everyone is current. By streamlining communication and collaboration, sales teams can work more efficiently, avoid miscommunications, and close deals faster.

Sales Analytics And Reporting; A CRM offers robust reporting and analytics capabilities to track sales performance, identify trends, and uncover areas for improvement. By analyzing these insights, sales teams can identify bottlenecks, assess which strategies work, and optimize their sales processes accordingly. This data-driven approach enables teams to make data-backed decisions that lead to shorter sales cycles and increased revenue generation.

Mobile Accessibility; Many CRMs offer mobile applications, enabling sales representatives to access customer information, update records, and track sales activities. Mobile accessibility empowers teams to be more responsive and agile, reducing the time it takes to provide information or resolve customer inquiries. With real-time access to data, sales representatives can engage with prospects and move them through the sales cycle more efficiently, speeding up the overall process.

Drive Data-Driven Decision-Making; Leaders often crave data-driven insights to make business decisions. A CRM serves as a centralized repository of customer data, providing leaders with real-time reports, analytics, and dashboards. By having access to accurate and up-to-date information, leaders can make strategic decisions based on factual analysis, leading to more efficient resource allocation, targeted investments, and improved financial results.

Customer Segmentation And Targeting; A CRM allows leaders to segment customers using various criteria such as demographics, purchase history, or engagement level. By analyzing these segments, leaders can identify high-value customer groups and tailor their marketing efforts to target them specifically. For instance, if the data shows that a particular demographic group has a higher average order value, leaders can allocate resources toward targeted marketing campaigns to attract more customers, resulting in improved financial results.

Lead Conversion Analysis; A CRM tracks the customer journey, from initial contact to conversion. With this data, leaders can analyze the effectiveness of different lead generation strategies and lead nurturing tactics. Leaders can allocate resources towards those strategies and optimize their sales and marketing efforts by understanding which channels or activities are most successful in converting leads into customers. This data-driven approach helps efficiently allocate resources and focus on activities that yield fantastic results.

Forecasting And Demand Planning; A CRM provides real-time data on sales trends, customer preferences, and purchasing patterns. Leaders can leverage this data to forecast demand, plan inventory levels, and optimize production or procurement processes. For example, suppose the CRM data indicates a surge in customer interest in a particular product. Leaders can adjust production schedules or stock levels to meet customer demand and avoid shortages or excess inventory. By using data-driven demand planning, leaders can optimize resource allocation and reduce costs, improving financial outcomes.

Customer Satisfaction Analysis; A CRM can capture and analyze customer feedback, reviews, and satisfaction ratings. Leaders can track key metrics like customer satisfaction scores or

Net Promoter Scores (NPS) to understand how well their products or services meet customer expectations. This data allows leaders to identify areas for improvement, make necessary adjustments, and enhance customer experiences. By prioritizing data-driven improvements to improve satisfaction, leaders can cultivate strong customer loyalty and generate more revenue from repeat business and referrals.

Marketing Campaign Performance Analysis; A CRM tracks and analyzes the performance of marketing campaigns across various channels. Leaders can gain insights into which campaigns are most effective in driving conversions and generating revenue. By analyzing these data-driven insights, leaders can refine their marketing strategies, optimize their marketing budgets, and focus resources on campaigns that have proven to deliver the best return on investment. This leads to more efficient resource allocation, targeted investments, and improved financial results.

By showcasing these specific ways a CRM can generate more revenue, you can persuade leaders that the cost is justified and necessary for long-term business success. Remember to emphasize the potential for cross-selling and upselling, customer retention, marketing optimization, streamlined sales processes, and data-driven decision-making.

28

The Cost Of A CRM - Investment Needs To Be Clarified

CRM is essential for companies seeking to maintain robust client relationships. Despite its many benefits, CRM can be challenging to implement correctly and often incur significant costs. Hence, it is crucial to understand how to assess these costs and ensure that your CRM implementation adds value to your organization.

One of the most critical factors to consider when evaluating a CRM system's cost is its potential impact on revenue. While CRM is only one factor influencing your company's income, assessing its potential impact on revenue is essential. You may need to dig deeper by monitoring progress via revenue per user, customer acquisition cost, or conversion rate to do so. However, it is worth noting that many CRMs are not directly tied to sales and may not significantly impact revenue. Consequently, evaluating the impact of a CRM implementation on income may be challenging and require in-depth analysis.

Moreover, connecting metrics to CRM actions is critical to assess a system's impact on revenue. This involves tracking key metrics after implementing the CRM and comparing them to the data collected before the implementation to determine if any changes occurred and if the CRM is responsible for them. For instance, if a sales CRM allows for increased dissemination volume and outreach platforms, monitoring revenue changes between the old and new systems is more accessible. However, it is essential to note that a design may boost one metric, such as sales call volume, without necessarily resulting in an immediate increase in profits. Thus, it's best to start with a proposition: the CRM will boost specific behaviors, resulting in specific desired outcomes. To achieve this, you will need a baseline of existing segmentation or sales data to compare against future sales data.

It is also prudent to consider the CRM's financial impact, as potential cost savings can be a significant driver of implementing a CRM system. Savings can come in various forms, such as less expensive bookkeeping, streamlined reporting, or custom reporting. Therefore, to assess a CRM's cost, you must determine the cost savings that come with its implementation, benchmark existing costs, have a cost-reviewing mechanism in place, and a metric to link the cost savings to the new CRM.

It is crucial to consider the many CRM implementation costs not immediately visible to make the most accurate cost-saving projections. These hidden costs may include user training, software integration, data migration, and ongoing maintenance. These aspects may lead to unforeseen expenses that can ultimately drive up the cost of implementing a CRM system.

Implementing a CRM system can significantly impact your organization, your customers, and all parties involved in the project. Therefore, it is vital to take a holistic approach to analyzing the costs and benefits of a CRM implementation. This will typically include consulting with all stakeholders to identify and evaluate potential hidden expenses and scope the project's size and potential impact.

29

Effect On Revenue

One crucial aspect that executives should consider when implementing a CRM system is its effect on revenue. **(To this point, the CRM team must understand these expectations and adjust as needed to meet them.**) Maximizing revenue is always the primary goal of businesses, and the CRM system's effectiveness should be carefully monitored to ensure that its implementation boosts the company's bottom line. The most apparent way to tie a CRM system's energy to revenue is to consider the company's financial metrics directly linked to revenue. However, a more in-depth analysis is required if such metrics are not directly tied to revenue.

A CRM system's effect on revenue is challenging to monitor because many factors are involved. Therefore, determining the appropriate metric to use requires carefully evaluating several factors. Identifying the proper metric is critical for the overall success of the CRM system implementation. It is often necessary to dig deeper and discover a metric that works best for your business, bearing in mind that what works best for one company may not work for another.

One way to determine a suitable metric to measure a CRM

system's effect on revenue is to evaluate the cost per customer acquisition or the conversion rate. By doing so, businesses can determine whether there was ultimately any improvement in revenue after the CRM system implementation. Calculating your average sales per customer for the period before you installed the CRM will give you a baseline for your data. After installing the CRM, you can calculate your average sales per customer, compare these numbers, and see whether there's an improvement. If there is an improvement, congratulations! You are on your way to realizing the full potential of your CRM system.

To measure the impact of a CRM system's implementation on revenue, companies should set up a baseline for their data before the implementation. It would be best if you also had some idea of how long it takes for a new customer to turn into a paid user (or whatever metric makes sense for your business). Doing so lets you track this number over time and see if any changes or improvements can be attributed to the CRM system's implementation.

It is important to note that no universal metric for determining a CRM system's effectiveness in generating revenue. The metric used should be aligned with the particular needs and objectives of the company, and it should be easy to track and compare. In cases where it is challenging to find a sales CRM that measures revenue, businesses can use alternative tools like HubSpot or Google Analytics. However, tracking revenue should be a priority for every business.

Ensuring the CRM system integrates seamlessly with the company's existing systems, databases, and applications is also critical. The integration enables data to flow smoothly across the company, and users can access relevant customer data in real time, providing better service to clients. The CRM system's flexibility is equally crucial, ensuring that features can be customized to meet the company's specific business requirements. The ability to accommodate future growth is essential, and the system should be scalable to meet possible future changes, remaining cost-effective at the same time.

Another point to consider when implementing a CRM system is

the impact on employees. CRM user adoption can be one of the biggest challenges companies face, and it is often the cause of failure in many CRM implementations. Employees need to be trained on how the system works and its benefits, and adequate support must be in place when encountering difficulties. Therefore, the system's ease of use is crucial, ensuring that navigating and understanding the system is intuitive and not overly complicated. The training process should be well-planned and implemented, and the CRM system's support framework should have the expertise to handle employees' concerns effectively.

Implementing a CRM system ranks as one of the most significant investments companies can make to improve CX and increase operational efficiencies. However, the implementation requires careful consideration and planning to generate the desired results. Identifying the appropriate metrics to monitor the CRM system's effect on revenue is critical for the project's success. The metric should be specific, easy to watch, and aligned with the business objectives. Integration with existing systems, flexibility, ease of use, and user adoption are critical factors. Ensuring adequate support for employees and the availability of expertise when they encounter challenges is also crucial. By adopting a careful and measured approach to implementing a CRM system, the company can fully maximize its benefits, position itself for sustainable growth, and achieve its revenue goals.

30

Case Study: The Wonky Allocation Game

Scene:

In a bustling corporate office, a group of project team members are gathered in a conference room for a meeting. The topic of discussion is the allocation and payment model for their company's modern CRM. The room is tense as everyone knows this issue has caused problems.

Incident:

The story unfolded with a disturbing incident involving a duplicitous divisional leader. This particular episode differed from previous issues that arose due to a lack of understanding or knowledge. This time, it was deliberate, sending shockwaves through the organization.

The divisional leader in question had proved to be a master

manipulator. They recognized the power and potential of the CRM system and chose to exploit it to their advantage. This individual maneuvered their way into positions of influence and control by skillfully navigating the system's intricacies. However, their actions came at a cost for others within the organization.

As the CRM leader continued their account, it became apparent that innocent parties were made to pay the price. As the duplicitous divisional leader manipulated the system, they distorted data, skewed results, and undermined the integrity of the CRM platform. The consequences rippled across departments, impacting decision-making, resource allocation, and trust in the CRM system.

The CRM leader emphasizes the importance of using a simple and fair payment model. He advises staying away from complex calculations tied to hiring and termination schedules, as these can create opportunities for manipulation. Instead, he suggests a model that charges everyone at the firm based on headcount rather than individual users — recalling an example from his previous company, where an allocation model was used that calculated costs once a year.

Unfortunately, the team responsible for system operations was disconnected from the team that distributed allocations, creating space for potential trickery. Several particular divisional leaders took advantage of this disconnect by decreasing their team's license count just before the allocation date, claiming they didn't see value in the system. As a result, he was charged significantly less than he should have been, while others had to pick up the slack.

When unused licenses are not paid for, the cost burden falls on the actual users. Those assigned licenses pay more per license because the pool of unused licenses increases. In the example, 50 licenses were returned, causing the cost per license to increase for those using them.

To make matters worse, months later, the same group requested the returned licenses again, exploiting the screwy allocation system to increase their bottom line.

Result:

The incident involved the complex chargeback model many companies use to allocate costs for their CRM. The CRM leader explains that if not carefully managed, this model can create loopholes that people can exploit. He describes how some individuals find ways to avoid paying their fair share, either by stating that they don't need the system, they are too small, returning licenses right before the allocation calculation period, or by piling up the workload on a few individuals to save on licenses.

Lessons Learned:

The incident highlights the unethical behavior of exploiting a flawed allocation system and the failure of a cowardly CFO to address the issue. Other leaders need to stress the need for consequences for such behavior and suggests tying the operators responsible for system allocation to those who actually use the licenses. This would create accountability and discourage bad behavior.

The critical lesson is to make the allocation or internal payment model simple and transparent. By tying the operators to the allocators and establishing consequences for unethical behavior, companies can prevent exploitation and create a fair and equitable system for all users.

31

Knowledge Management

CRM and Knowledge Management (KM) are considered separate disciplines with distinct roles and goals. And they are. But with the flow of time, it became clear that they both have overlapping and complementary purposes and must work together to provide continual improvement to company clients. Especially that the cutthroat environment of today has refocused firms' attention on customer happiness.

CRM and KM have become critical strategic tools for all businesses in today's world of fierce competition. They both stress the importance of allocating resources to business-supporting activities to achieve a competitive advantage.

Given the importance of KM systems in today's customer-centric corporate environment, a comprehensive and straightforward framework to incorporate traditional CRM functionalities with the planning and using customer-related expertise, specifically in business choices, is lacking. Traditional CRM has concentrated on transactional exchanges to manage customer interactions. Still, KM systems handle an organization's information by developing, organizing, distributing, and utilizing knowledge to improve performance and create value.

As businesses grapple with the unprecedented complexities and challenges of modern markets, KM has emerged as a critical component of contemporary organizational strategies. With the sheer volume of information available in the digital age, enterprises must generate, curate, and disseminate knowledge to meet a growing array of needs. Therefore, KM has emerged as a critical element, facilitating the efficient management and effective use of knowledge resources across the enterprise.

In practice, KM's applications manifest in the following areas:

Knowledge-Based Accessibility And Upkeep; This represents the first phase of KM. It involves designing, organizing, and granting access to a body of knowledge. While this area may seem deceptively simple, it represents a vital and continuous challenge. Failure to manage information effectively creates tangible risks that can severely impact business agility and performance. Access to timely and relevant information is essential to support effective decision-making and action execution and can spell the difference between success and failure.

Expertise Management; The second area concerns extracting intelligent and relevant knowledge from static data by enabling experts to process and understand the data. This expertise-derived intelligence can take many forms, including recommendations generated through machine learning algorithms. However, an enterprise's experts are vital to effective KM, given their unique insights and perspectives in various domains. Identifying and leveraging these experts' talents can significantly benefit the organization, reducing errors, improving quality, and generating new ideas.

Collaboration; Facilitating the generation of new information represents a third critical area of KM. Collaboration tools are essential to this area, fostering group interaction and allowing them to work on more extensive and complex projects. Collaboration elicits greater participation from stakeholders, increasing knowledge sharing. Collaboration tools promote synchronous and asynchronous interactions, providing flexibility to work across time zones and regions. Collaborative systems also allow users to capture the tacit knowledge of their teams, translate this knowledge into new forms and share it with other groups, thereby facilitating cross-functional collaboration.

Business Applications For KM; The final area of KM comprises applications to support knowledge-based activities such as e-learning and business analytics. In our data-driven world, knowledge analytics is coming of age, where enterprises analyze and extract insights from customer interactions, social media, and data streams to drive performance. The wisdom available through analytics generates innovative and informed business decisions leading to a competitive edge and agility. Knowledge-driven analytics can be applied across many industries, such as healthcare, banking, retail, or finance, focusing on customer engagement, product development, and process improvements.

IT IS essential to recognize that KM is a hugely important, serious, professional, and ongoing and continuous process - it is not a one-time event that can resolve all issues. The applications of KM have evolved significantly, changing the face of enterprise capabilities and driving tremendous success and agility. Every member of an organization can benefit from KM applications, making it essential to identify and leverage capabilities effectively. Stakeholders must also recognize that effective KM is built on a robust governance, accountability, and continuous learning foundation. Enterprises that frame their approach to KM in this fashion and establish formality around their KM efforts have a higher probability of success. By seeking to manage knowledge strategically and using technology to accelerate these efforts, KM practitioners can help organizations identify and leverage knowledge more effectively, enhancing performance and competitive advantage.

The case for KM and CRM integration is becoming more compelling by the day. Customers nowadays want better quality products and faster service that is more individualized. By itself, CRM cannot provide tailored services on such a large scale.

When customer knowledge and CRM are combined, businesses prioritize their clients. E-commerce businesses have developed a dynamic buying and selling strategy to ensure that customers get the best possible outcomes from a mix of suppliers, services, and procedures. This applies to both B2C and B2B clients.

Blending KM and CRM to work together seamlessly is a strategic move that can unlock the full potential of your organization, providing you with a long-term competitive advantage. By combining customer insight with corporate knowledge, enterprises can significantly improve productivity, efficiency, customer satisfaction, agility, and consistency. However, it is crucial to approach this integration with caution and due diligence, as some potential challenges and considerations must be addressed to ensure successful implementation and maximum benefits for all stakeholders involved.

One of the immediate benefits of KM and CRM integration is improved productivity. Employees can access a wealth of information and answers to commonly asked questions directly from the knowledge management system while simultaneously addressing customer inquiries in CRM. This eliminates the need for time-consuming information searches, conversations with colleagues, and program switching. By streamlining these tasks, employees can respond more quickly and effectively to customer needs, reducing wait times and enhancing overall productivity.

In addition to productivity gains, integrating KM and CRM enables improved efficiency. By tracking the utilization of the knowledge base in addressing calls, emails, and inquiries in real-time, organizations can monitor the efficacy and productivity of front-end staff. This real-time accountability ensures that the knowledge base consistently provides accurate and valuable information to support customer interactions. By identifying areas of improvement and refining the knowledge base, organizations can continually enhance efficiency and ensure that customer inquiries are addressed more effectively.

Furthermore, the integration of KM and CRM reduces unnecessary tasks by continuously updating the CRM system with information gathered from each customer interaction. By documenting answers to common issues and saving them as valuable skills within the knowledge base, organizations create a growing repository of knowledge that can benefit the entire company. This reduces the need for repetitive work and redundant inquiries and empowers employees with the most up-to-date information, enabling them to respond quicker and more accurately to customer queries.

Enhanced customer satisfaction is another critical advantage of KM and CRM integration. A company's website visitors can access the knowledge management system and digital FAQs to independently identify and resolve their issues. Organizations empower customers to find relevant solutions conveniently by providing a comprehensive self-service option, leading to higher satisfaction levels. This self-service option also reduces customer support burnout, allowing them to focus on advanced customer inquiries and provide personalized assistance where needed.

In today's fast-paced business environment, agility is crucial for success. By combining KM and CRM, organizations can obtain valuable customer input and feedback more rapidly, enabling them to iterate on products and services more efficiently. This shortened feedback loop accelerates the time to market for new offerings, providing a competitive edge and capitalizing on emerging opportunities. The integration of KM and CRM facilitates capturing and utilizing customer insights, ensuring that decision-making is informed by real-time data and fostering a culture of continuous improvement.

One often overlooked benefit of KM and CRM integration is consistency. Establishing a centralized knowledge repository ensures that information is uniform and consistent across all customer touch-points. In a dispersed system where documents addressing the same themes exist in multiple locations, inconsistencies can arise, leading to customer confusion and frustration. Users expect reliable and up-to-date information, and any deviations can result in a negative experience. KM's central knowledge management database resolves this issue by providing a single "book of record", allowing customers to access consistent and reliable information across all interactions.

Delivering exceptional customer service in today's complex and competitive climate is undoubtedly challenging. However, many organizations recognize knowledge management's power in optimizing staff performance and motivating employees. Companies can significantly improve service delivery speed, precision, reliability, and adaptability by embracing knowledge management as a core customer service strategy. Accessing critical information efficiently and leveraging collective knowledge empowers employees to provide exceptional customer experiences, ultimately leading to customer loyalty and business growth.

Integrating KM and CRM offers substantial benefits for organizations seeking a competitive advantage in the long term. By combining customer insight with corporate knowledge, companies can improve productivity, efficiency, customer satisfaction, agility, and consistency. However, it is crucial to approach this integration with caution, implementing proper governance structures and ensuring ongoing maintenance and updates to the knowledge management system. Organizations prioritizing knowledge management as a core component of their customer service strategy will be better equipped to meet the needs of a dynamic marketplace and exceed customer expectations.

That's A Wrap

You can stop reading now. Class dismissed! I hope you found value in reading this book, and perhaps it even triggered some hidden gems of wisdom tucked away in your mind's recesses. If that's the case, I apologize for potentially unsettling your thoughts. Still, a gentle nudge is sometimes necessary to bring those insights to the surface. On the other hand, it's equally possible that you had already recognized these lessons all along but required the validation of seeing them in writing. In that case, you're welcome. Validating our instincts and experiences with the wisdom of others can be a powerful tool for growth and decision-making. It was a fun process, and I surprised myself as I didn't know I had so much to say - **who knew?**

www.ingramcontent.com/pod-product-compliance
Lightning Source LLC
LaVergne TN
LVHW050906210926
848482LV00012B/357
* 9 7 8 1 7 3 6 0 2 1 3 6 1 *